Solid State Devices — A Quantum Physics Approach

Other Macmillan titles of related interest

J. C. Cluley, *Electronic Equipment Reliability*, second edition
R. F. W. Coates, *Modern Communication Systems*, second edition
M. E. Goodge, *Semiconductor Device Technology*
B. A. Gregory, *An Introduction to Electrical Instrumentation and Measurement Systems*, second edition
Paul A. Lynn, *An Introduction to the Analysis and Processing of Signals*, second edition
Paul A. Lynn, *Electronic Signals and Systems*
A. G. Martin and F. W. Stephenson, *Linear Microelectronic Systems*
J. E. Parton, S. J. T. Owen and M. S. Raven, *Applied Electromagnetics*, second edition
A. Potton, *An Introduction to Digital Logic*
J. T. Wallmark and L. G. Carlstedt, *Field Effect Transistors in Integrated Circuits*
G. Williams, *An Introduction to Electrical Circuit Theory*

Solid State Devices
A Quantum Physics Approach

Donard de Cogan

*Department of Electrical
& Electronic Engineering
University of Nottingham*

MACMILLAN
EDUCATION

© Donard de Cogan 1987

All rights reserved. No reproduction, copy or transmission
of this publication may be made without written permission.

No paragraph of this publication may be reproduced, copied
or transmitted save with written permission or in accordance
with the provisions of the Copyright Act 1956 (as amended).

Any person who does any unauthorised act in relation to
this publication may be liable to criminal prosecution and
civil claims for damages.

First published 1987

Published by
MACMILLAN EDUCATION LTD
Houndmills, Basingstoke, Hampshire RG21 2XS
and London
Companies and representatives
throughout the world

Typeset by TecSet Ltd, Wallington, Surrey
Printed in Great Britain by
Camelot Press Ltd,
Southampton

British Library Cataloguing in Publication Data
De Cogan, Donard
   Solid state devices: a quantum physics
   approach.
   1. Solid state physics
   I. Title
   530.4'1   QC176

ISBN 0-333-40972-8
ISBN 0-333-40983-6 Pbk

To
Anne and Dominic

Contents

Preface		*x*
Symbols used for Variables		*xii*
1	**The Elements of Crystallography**	**1**
	The crystalline state	1
	Close packing and crystal structure	3
	Directions within a crystal	4
	Distances between planes in a simple cubic crystal	5
	X-ray crystallography	7
	Further developments in X-ray crystallography: the reciprocal lattice	12
	Problems	14
	Reference	15
2	**Matter Waves**	**16**
	The development of quantum mechanics	16
	The duality concept	18
	de Broglie's hypothesis	18
	Fourier's theorem	21
	Heisenberg's Uncertainty Principle	24
	Problems	26
	References	27
3	**Wave Mechanics**	**28**
	Introduction	28
	Review of wave motion	28
	Wave functions	30
	Observables and operators	32
	Wave mechanics	35
	Boundary conditions for solution of Schrödinger's equation	36
	Problems	49
	Reference	51
	Further Reading	51

4 Quantum Theories of Solids — 52

The Free Electron Theory of Solids — 53
Fermi–Dirac statistics — 54
The derivation of $S(E)$ — 55
Thermionic emission — 58
Field-enhanced emission — 60
Field emission (Fowler–Nordheim tunnelling) — 62
Photo-electric effect — 64
Band Theory of Solids — 64
The velocity of an electron in an energy band — 72
The effective mass of an electron in an energy band — 72
Problems — 74
References — 76

5 Electrons and Holes in Semiconductors — 77

The position of the Fermi level in an intrinsic semiconductor — 78
The position of the Fermi level in an extrinsic semiconductor — 79
Carrier concentrations in semiconductors — 80
Majority and minority carriers — 82
The motion of carriers in semiconductors — 83
Semiconductors in non-equilibrium — 86
Carrier generation and recombination — 87
The steady-state condition — 88
Return to equilibrium — 88
Steady-state injection from a boundary — 89
Problems — 91
Reference — 92

6 p–n Junctions — 93

Position of the Fermi level at equilibrium — 93
The amount of band bending at equilibrium — 94
The width of the depletion layer as a function of doping densities — 95
Application of bias — 97
Capacitance behaviour of p–n junctions — 98
Current transport in p–n junctions — 99
Junction breakdown under reverse bias — 103
Temperature dependence of junction breakdown — 104
Real diodes — 105
Problems — 106
Reference — 107

CONTENTS ix

7 Junction Transistors — 108

Unipolar (field effect) transistors — 108
Bipolar transistors — 111
Transistor gain — 115
Non-ideal injection efficiency — 115
Some secondary effects on bipolar transistor performance — 116
Some considerations for good bipolar transistor design — 117
Real bipolar transistors — 117
Problems — 118
Reference — 119

8 Surface Effects and Surface Devices — 120

The metal-semiconductor contact — 120
The metal-oxide-semiconductor (MOS) contact — 122
Metal-oxide-semiconductor (MOS) devices — 127
Conclusion — 132
Problems — 132
Reference — 134

Solutions to Problems — *135*

Physical Constants and Conversions from Non-SI Units — *150*

Index — *151*

Preface

The changes which have taken place in electronics are truly astonishing. It is difficult to believe that within a single lifespan we have come from the cat's-whisker diode, via the thermionic valve, to the 256K random access memory and beyond. These developments would not have come about without an increased understanding of the physics and technology of the solid state.

Although the progression from Planck's quantum postulate to single chip electronic systems within eighty years has resulted in an increased level of specialisation of the fields of knowledge, solid state nevertheless continues to be a cross-disciplinary subject. The design and fabrication of solid state devices involve large elements of chemistry, physics and materials science. However, books on the subject tend to be written by specialists in one or other area. Thus a physics-based text is likely to have more details on quantum theory than is necessary for a technologist. Similarly, texts which concentrate on devices and their applications frequently ignore the fundamental background which is vital for a true understanding.

This book attempts to bridge the gap by presenting a clear résumé of the background to semiconductor device theory in a multidisciplinary way. It has its origins in a course which has been taught in Electrical and Electronic Engineering at Nottingham University for several years. This was intended to give students a simple grounding in quantum theory and wave mechanics in order to provide the conceptual and manipulative skills for subsequent courses in integrated circuit technology, VLSI and power semiconductor electronics. Although the book follows the essential outlines of the course, it has been adapted for a wider readership who will nevertheless find it beneficial to have some background in physics and mathematics – probably equivalent to that required for entry to an undergraduate course in science or engineering.

Crystallography is used as the starting point. There is an introduction to the reciprocal lattice and the application of Fourier's theorem to X-ray structural analysis. The use of Fourier methods is extended in the next chapter to provide a building block for wave mechanics. The implications of wave mechanics are not far removed from concepts which students might well encounter in an electromagnetics course. Basic ideas which are common in the area of molecular bonding theory are used to lead the reader towards the solid state. The solution of Schrödinger's equation for extended problems is treated in terms of the Free Electron and Band theories. The introduction of the reciprocal lattice in the first chapter should assist with an understanding of k-space.

Once a quantum mechanical foundation for solids has been provided, attention is then focussed on semiconducting materials and semiconductor devices. Further background includes treatments of diffusion, drift and recombination processes which are used when considering two-terminal, three-terminal unipolar and three-terminal bipolar structures. The final chapter on the properties of semiconductor surfaces and semiconductor surface devices provides an introduction to MOS, with some pointers towards MOS integrated circuits.

Space constraints have limited any detailed discussion of optical and thermal properties to implicit treatments within the framework given above. For similar reasons, mention of magnetic properties has had to be omitted.

Although the level is not particularly sophisticated, there has been no dilution of mathematical content. I have attempted to avoid the dreaded phrase "it can easily be seen that . . ." and all derivations are given in full. Where relevant, references for background reading have been provided. These appear at the end of chapters along with a set of problems. The problems have been designed with several aims in mind, the most important being the consolidation of comprehension and the conceptual extension of text material. Solutions are provided for all numerical examples.

In the preparation of this book I would like to thank Professor B. Tuck and the pre-publication referees whose comments have helped in no small way to convert a set of course notes into a broad-based text. I would also like to thank the succession of Nottingham students who have, by means which were sometimes diplomatic, sometimes less so, forced a gradual evolution of which this is the end product.

Finally, few authors can be so lucky as to have an artist for a father. I would like to thank mine for the many hours which he has devoted to the illustrations.

Symbols used for Variables

C	capacitance
d	Cartesian space distance
d^*	reciprocal space distance
D	diffusion constant
$\mathscr{E}$	electric field
E	energy (for example, E_f = Fermi energy, E_g = band gap energy)
$[E]$	total energy or Hamiltonian operator
f	frequency
F	flux (for example, F_p, F_n = hole and electron flux)
$[H]$	alternative symbol for total energy or Hamiltonian operator
k	wave number
L	diffusion length (for example, L_p, L_n = hole and electron diffusion length)
m	mass
m^*	effective mass
n	electron concentration (for example, n_p = electron concentration in p-type semiconductor)
N	impurity concentration (for example, N_A, N_d = acceptor and donor concentrations)
p	hole concentration (for example, p_{n_0} = hole minority carrier concentration at equilibrium)
P	momentum
$[P]$	momentum operator
Q	electrical charge
R_H	Hall coefficient
$S(E)$	density of electron states
T	temperature
$U_k(x)$	block function
v	velocity
V	potential
W	depletion layer width
W_b	base width of a bipolar transistor
x	position in Cartesian space

γ	injection efficiency
ϵ	permittivity
λ	wavelength
μ	mobility
ρ	electrical resistivity
σ	standard deviation of a normal distribution
τ	minority carrier lifetime (for example, τ_p, τ_n = hole and electron lifetimes)
ϕ	work function
ϕ_B	built-in potential of a p-n or metal/semiconductor junction
ψ	wave function
ψ^*	conjugate of wave function
ω	angular velocity

1

The Elements of Crystallography

The crystalline state

In a gaseous assembly of atoms, ions or molecules at high temperature, the interactions between components of the system are negligible compared with the thermal energy, kT. However, as the temperature is reduced the average separation between components decreases and interactions become more significant. Two phase/energy transitions normally occur at successively lower temperatures. The latent heat of vaporisation is emitted when the gas condenses to a liquid. The liquid state is characterised by short-range order, but long-range disorder. The latent heat of fusion is emitted when the liquid/solid transition occurs. In the solid state the intercomponent separations are very small, and long-range interactions are possible.

If a liquid/solid phase change occurs very quickly, as with the rapid cooling of a liquid, an amorphous solid is obtained. This is sometimes called the 'glass state' and is characterised by an apparent lack of order in the arrangements of the components. If, on the other hand, the rate of decrease of temperature is sufficiently low as to permit the components of the liquid to adopt energetically favourable positions relative to their neighbours, then a crystalline solid is formed. The crystalline solid, which is characterised by a high degree of order and symmetry in the arrangement of its components, has a lower energy than an amorphous solid at the same temperature. Accordingly, glass is said to be in a state of metastable equilibrium and devitrification may occur as a result of the transition from the metastable (amorphous) to the stable (crystalline) state.

There are two main types of crystalline solid. Ionic solids have components which have electrostatic charges. There are no electrostatic charges on the components of a covalent solid. In reality, many solids lie somewhere between these two extremes. Covalent solids can be atomic as in a metal, compound as in gallium arsenide or cadmium sulphide, or molecular as in anthracene or polyethylene.

At first sight it is not very easy to see what holds a covalent crystal together, but this should become clearer during the course of this book. On the other hand, ionic crystals can be very successfully treated using a model which considers electrostatic attractions and repulsions based on Coulomb's law. In sodium chloride (NaCl), each sodium cation (Na^+) is surrounded by several chloride anions (Cl^-). These anions in turn attract further cations which attract anions, and so on. The arrangement of ions is such that there is a net drop in electrostatic energy relative to an equal quantity of gaseous ions held infinitely far from each other. This increase in stabilisation when a solid is formed is called the *lattice energy*. It can be calculated using either a thermodynamic approach called the Born–Haber cycle or an electrostatic approach called the Born–Madelung method. Both of these methods are covered in more detail in Moore's book *Seven Solid States* (Moore, 1967).

In addition to the net attractions which hold a solid together, there are also short-range repulsions which ensure that they do not come too close to each other. Because of the two opposing forces, there is an equilibrium distance between any two atoms or ions. Thus, each component of a crystal has a definite radius which represents half the closest distance of approach for a similar component. Some crystal radii are given below.

Si	1.175 Å	Au	1.442 Å
Ge	1.225 Å	Na^+	0.970 Å
P	1.090 Å	Cl^-	1.810 Å
Ga	1.220 Å	O^{2-}	1.320 Å

Because each atom or ion in a crystal has a definite radius, it is convenient to treat the components of a crystal as a collection of billiard balls. This analogy

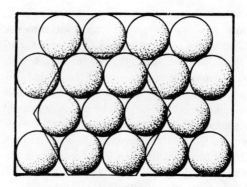

Figure 1.1 A close pack arrangement of billiard balls.

can be used in a treatment of crystal symmetry and crystal forms. The approach used here will be in terms of a close packing arrangement of atoms.

Close packing and crystal structures

If one starts with a large collection of billard balls of equal size and places them in a tray, they will close-pack as shown in figure 1.1. Taking a small portion of the tray, we can recognise some salient features. There are many voids similar to that marked B in figure 1.2, where there is a curved triangle with its apex pointing upwards. There are also voids, marked C, where the apex points downwards.

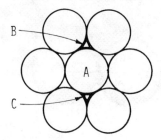

Figure 1.2 A detail of the close pack showing the B and C sites.

When one comes to fill the tray with a second layer of billiard balls one can place them on either B voids or C voids, but not both if one is to get a true close packing. It is worth remembering that balls placed on C voids are identical to those placed on B voids after a 180° rotation of the tray. Filling the third layer allows a certain choice. Assuming that the second layer was on B voids, the third layer could be placed on either C or back on A so that it copies the first layer. If C is the choice for the third layer and if the process is repeated, one gets a sequence ABCABC, as shown in figure 1.3. This packing gives a cubic arrange-

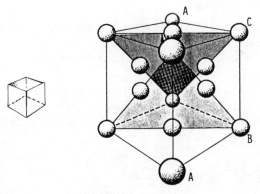

Figure 1.3 An ABCABC ... arrangement of close packing of successive layers. The result is a face centred cubic structure.

ment with a ball at the centre of each face. Accordingly, it is called a face centred cubic (fcc) close packing.

An ABAB ... arrangement of balls (figure 1.4) gives a hexagonal close packing.

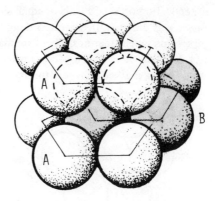

Figure 1.4 An ABAB ... arrangement of close packing of successive layers. The result is a hexagonal close packed structure.

The ABCABC (fcc) close packing is only one example of cubic structure. There is also the simple cubic structure, which has a ball at each corner. Caesium chloride is an example of a body centred cubic structure (bcc) where a Cs^+ ion sits in the centre of a cube and is surrounded by Cl^- ions, one at each of the eight corners. There are many other crystal structures but the cubic arrangements mentioned above are sufficient for the discussions which follow.

Directions within a crystal

A plane of atoms can be described in space by specifying the points where it intersects an orthogonal x, y, z axis. However, there are problems when a plane is parallel to one or more of the axes. The plane shown in figure 1.5 would be $[\infty, 2, \infty]$.

This problem can be alleviated by using Miller indices which are the reciprocals of the x, y, z axis intersections normalised to 1. Thus the plane intersecting at $x = 2, y = 2, z = \infty$ has a Miller index of (1, 1, 0), usually written as (110); that is

$$\frac{1}{2}, \frac{1}{2}, \frac{1}{\infty} = (0.5, 0.5, 0) = (1, 1, 0)$$

There is a convention that Miller indices are bounded by round brackets to distinguish them from positional indices which are bounded by square brackets.

THE ELEMENTS OF CRYSTALLOGRAPHY

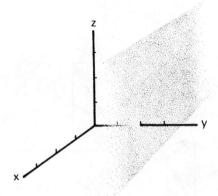

Figure 1.5 A plane which intersects the y axis at 2 and the x and z axes at infinity.

Note A Miller index specifies the direction of a plane. It cannot distinguish between parallel planes. $[2, 2, \infty]$ and $[4, 4, \infty]$ are both (110) planes.

Distances between planes in a simple cubic crystal

The distance between (100) planes in a simple cubic structure is equal to the separation between nearest atoms/ions (figure 1.6). The distances between adjacent (110) and (111) planes are shown in figures 1.7 and 1.8 respectively.

In general the distance between adjacent planes described by the Miller index (h, k, l) is

$$d_{hkl} = \frac{\sqrt{h^2 + k^2 + l^2}}{h^2 + k^2 + l^2} \times a$$

where a is the unit cube length.

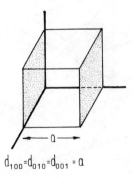

$d_{100} = d_{010} = d_{001} = a$

Figure 1.6 Separations between (100), (001) and (010) planes in a cube of length a.

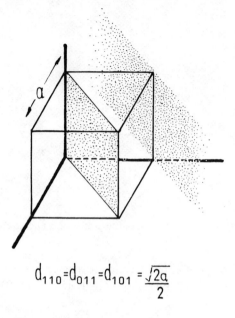

$$d_{110} = d_{011} = d_{101} = \frac{\sqrt{2}a}{2}$$

Figure 1.7 Separations between (110), (011) and (101) planes in a cube of length a.

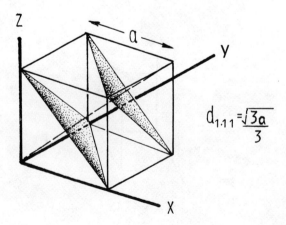

$$d_{1 \cdot 1 \cdot 1} = \frac{\sqrt{3}a}{3}$$

Figure 1.8 Separations between (111) planes in a cube of length a.

X-ray crystallography

The Bragg diffraction formula

It is well known that if two waves encounter each other they will interfere constructively if they are in phase (in step). If they are out of phase they will interfere destructively (cancel each other). This is the phenomenon of *diffraction*. It can be observed in any part of the electromagnetic spectrum.

The interference of light which is observed with a film of oil on the surface of water is one example of diffraction. Light waves are reflected at the oil-air interface. Other light waves penetrate the film and are refracted. A component of this light will be reflected at the oil-water interface. Some will be transmitted back into the air, while the rest is reflected within the film. The light emerging from the oil may be in or out of phase with the light reflected at the top surface. The phase difference depends on the additional distance travelled by the light which penetrated the oil. This, of course, depends on the thickness of the film. The semiconductor industry regularly uses this phenomenon as a rapid means of determining the thickness of the oxide films which are grown on silicon.

Nobel prize winners William and Lawrence Bragg (father and son) first suggested the application of this technique to the analysis of crystal structure.

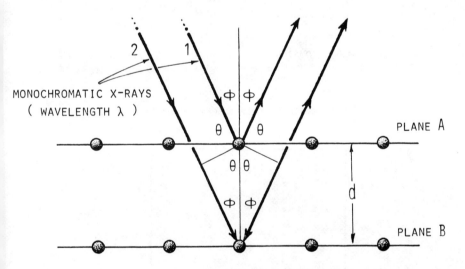

Figure 1.9 Scattering of X-rays from crystal planes.

Ray 2 in figure 1.9 travels a distance $2d \sin \theta$ further than ray 1. Constructive interference occurs if

$$2d \sin \theta = n\lambda$$

where $n = 1, 2 \ldots$ is the order of the diffraction, and λ is the wavelength. This is called the *Bragg diffraction formula*. Since d in a crystal is of the order of a few Ångstroms, λ must be of the same order if diffraction is to be observed. λ falls within the X-ray region of the electromagnetic spectrum and for this reason the technique is referred to as *X-ray crystallography*.

An X-ray diffraction analysis can be undertaken on either a single crystal sample, as in the Laue method, or on a sample in powder form, as in the Debye-Scherrer method.

Laue method

If a collimated beam of polychromatic X-rays is passed through a series of sodium chloride crystals, the Bragg condition ensures that a monochromatic beam is obtained. If this is directed on to a single crystal, as in figure 1.10, the

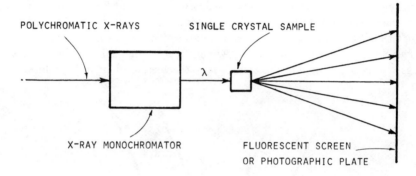

Figure 1.10 Schematic arrangement of a Laue diffractometer.

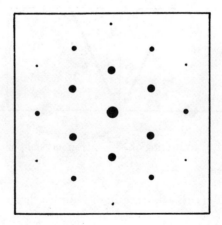

Figure 1.11 Laue picture of the (111) face of an iron crystal.

result is a regular arrangement of spots on the screen or photographic plate. The arrangement is a function of the crystal symmetry and angle of orientation of the sample to the beam (see figure 1.11). In order to obtain as much information as possible about the crystal, it must be examined in different orientations.

The Debye-Scherrer method

This method operates on the principle that a powder consists of a very large collection of microcrystals which present themselves with all possible orientations to the electron beam. If a given plane in a crystal [such as (111)] is presented to the beam with all possible orientations, the Bragg condition ensures that the locus of diffracted rays will describe a pair of cones apex to apex. If the pair of cones shown in figure 1.12 corresponds to first-order diffraction, there are similar pairs for $n = 2$ and so on. There will also be pairs of cones for other planes.

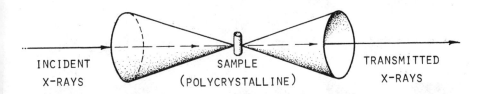

Figure 1.12 A single pair of diffraction cones.

A thin strip of film is placed round the radius of a Debye-Scherrer camera and the sample is placed at the centre (see figure 1.13a). The resulting pictures for sodium chloride and potassium chloride are shown in figure 1.13b. The pictures give the positions and relative intensity of the lines. From the radius of the camera and the position of a line relative to the centre of the film, the angle relative to the incident beam can be calculated. In a modern Debye-Scherrer camera the sample is placed in a stationary holder. Both the source and the detector (electronic rather than photographic) are moved through an angle θ.

In the older designs the powder sample was placed in a thin-walled quartz tube. In addition to being a means of support, the tube material provided a calibration. The diffraction lines for quartz (silicon dioxide) were always at the same position, independent of the sample under investigation.

10 SOLID STATE DEVICES – A QUANTUM PHYSICS APPROACH

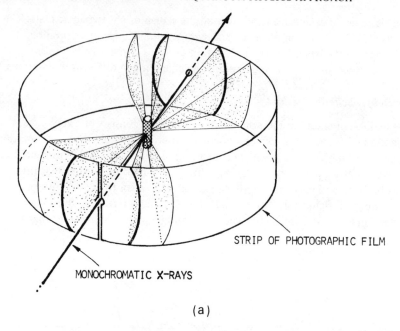

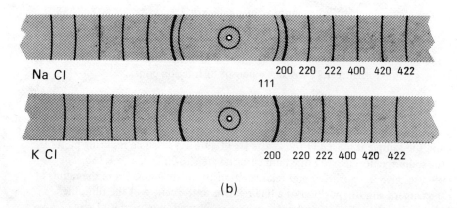

Figure 1.13 (a) Schematic arrangement of a Debye-Scherrer camera. (b) Debye-Scherrer patterns for sodium chloride and potassium chloride.

Example: Use of Laue diffraction to confirm the crystal structure of common salt

A single crystal of salt was examined using the K_α line of palladium (0.19 Å). First, second and third order maxima from the (100) planes were measured at

5.9°, 11.85° and 18.15°, which gives the (100) plane spacing as $d_{100} = 0.92$ Å. The first-order diffraction maxima for (110) planes was at 8.4° and for (111) at 5.2°. For constant λ and n the following ratios hold true

$$d_{100} : d_{110} : d_{111} = (\sin 5.9°)^{-1} : (\sin 8.4°)^{-1} : (\sin 5.2°)^{-1}$$

This ratio is approximately equivalent to

$$1 : 1/\sqrt{2} : 2/\sqrt{3}$$

which is only true for a face centred cubic structure, which must therefore be the structure of sodium chloride.

If the X-ray diffraction patterns are examined in more detail, some interesting points can be observed. The first and third order maxima from the (111) planes are weak, while the second and fourth order are strong. This is due to the relative size of the Na^+ and Cl^- ions. If one looks at the (111) planes shown in figure 1.14, it will be seen that the first contains only Cl^- ions which have a large radius — that is, 1.81 Å. The next plane contains only Na^+ ions which have a much smaller radius, 0.95 Å, and are therefore not such good reflectors. Most X-rays are scattered from Cl^- planes which explains why the $d_{100} : d_{110} : d_{111}$ ratio is $1 : 1/\sqrt{2} : 2/\sqrt{3}$ and not $1 : 1/\sqrt{2} : 1/\sqrt{3}$ which might be expected from geometric consideration of a simple cubic lattice. This effect is not so noticeable in potassium chloride (KCl) where the K^+ radius is 1.33 Å. Thus, it can be seen that not only can X-ray diffraction give details of the lattice spacing, but it can also confirm the crystal structure and give information about the relative sizes of ions in the crystal.

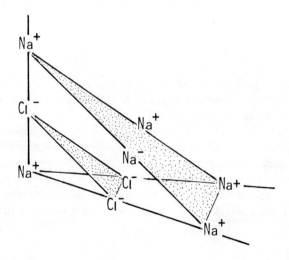

Figure 1.14 Successive (111) planes of sodium chloride.

Further developments in X-ray crystallography: the reciprocal lattice

In any formal treatment of crystallography it is necessary to devise some method of representing the planar spacing d_{hkl}. The reciprocal lattice is a very elegant way of approaching the problem. It has several distinct advantages. It can be treated numerically and is consistent with the concept of Miller indices. Remember, a plane that intersects the x, y, z axes at a, b, c has Miller indices $(1/a, 1/b, 1/c)$.

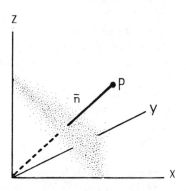

Figure 1.15 Reciprocal point P corresponding to a plane (hkl); $\bar{n}$ is the unit vector normal to the plane.

A family of planes described by the Miller indices (h, k, l) is represented by a vector **P** whose location is given by

$$d^{*}_{hkl} = \frac{1}{d_{hkl}} \bar{n}$$

$\bar{n}$ is the unit vector normal to the plane. Every plane in a crystal can be represented by a point P_{hkl}. The set of all points P_{hkl} forms a new entity called the reciprocal lattice, as shown in figure 1.16.

Figure 1.17 shows how a reciprocal lattice might respond to an incident beam of X-rays. As a crystal plane, (hkl), is rotated, the reciprocal lattice is also rotated and all reciprocal points within a distance $2/\lambda$ of the origin are brought into coincidence with the circumference of the circle of reflection. The scattering angle θ then depends on the distance between the origin and the reciprocal lattice point on the circle, which is given by an alternative form of the Bragg equation

$$\frac{1}{d_{hkl}} = \frac{2 \sin \theta}{\lambda}$$

THE ELEMENTS OF CRYSTALLOGRAPHY 13

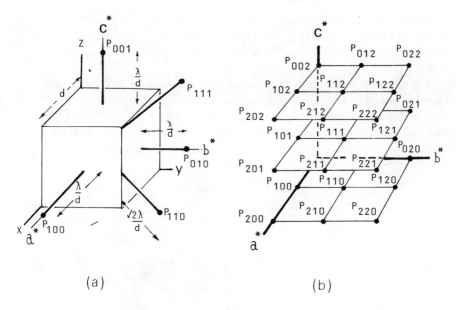

Figure 1.16 (a) Reciprocal points for several planes of a simple cube: the a^*, b^*, c^* reciprocal axis is shown. (b) A reciprocal lattice for a simple cube.

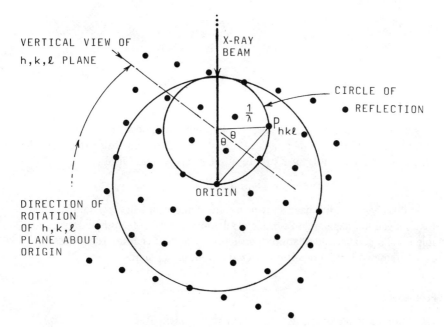

Figure 1.17 Ewald's construction of the Bragg diffraction condition.

Thus a Laue X-ray diffractograph is essentially a slightly distorted view of a single plane of the reciprocal lattice. Since it is possible to index each point in a reciprocal lattice, it is also possible by inspection to ascribe each Laue point to a crystal plane.

Any diffraction of X-rays can be traced back to the source of that diffraction using the principles of Fourier analysis which are presented in outline in the next chapter. It is the electrons associated with the atoms of a crystal which, in fact, reflect the incident X-rays. Thus a Fourier analysis using reciprocal lattice techniques can be used to compute the structure of a crystal in the form of an electron density distribution, as shown in figure 1.18.

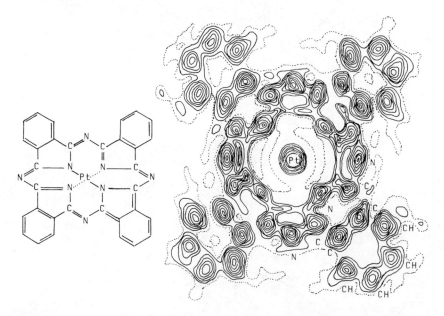

Figure 1.18 Electron density map of platinum phthalocyanine. It can be seen that the C-N bonds next to the metal are more distorted than the ideal molecular structure (see insert) would suggest.

In this chapter it has been shown that the components in a crystalline material are arranged in a regular and symmetric manner. Analytic techniques are available for the detailed elucidation of structure. The next chapter establishes an initial framework for a quantum mechanical theory of solids.

Problems

1.1. Aluminium has a density of 2700 kg m^{-3} and an atomic weight of 26.9.
   (i) Calculate the number of atoms per cubic metre.

THE ELEMENTS OF CRYSTALLOGRAPHY

(ii) If d_{100} for this face centred cubic (fcc) metal is 4 Å, calculate d_{110} and d_{111}.

1.2. A cube has several symmetry axes. An axis through the centre of opposite faces is said to have four-fold symmetry. This is because a 90° rotation about this axis leaves the cube appearing exactly as it was before it was rotated. An axis through the centre of diagonally opposite edges has two-fold symmetry.

What is the separation between planes which are normal to the three-fold axis of symmetry in a face centred cubic structure whose unit cell length is 4 Å.

1.3. Calculate the first-order diffraction angle which would be expected if a beam of X-rays (λ = 1 Å) was directed on to the (100) planes of the cube in the previous problem.

1.4. Sodium (density 970 kg m^{-3}, atomic weight 22.99) has a body centred cubic structure (d_{100} = 4.28 Å). Assuming that the atoms can be represented by rigid spheres, calculate
   (i) the ratio of the volume of the atoms to the total volume of the unit cube;
   (ii) the number of atoms per square metre on the (110) and (111) planes.

1.5. An ionic compound AB has a simple cubic crystal structure. It has a molecular weight of 58.45 and a density of 2177 kg m^{-3}. A beam of 1.54 Å X-rays is directed on to the (100) face. If a photographic plate were placed 10 cm in front of the crystal, what would be the separation between the first and second order reflections in the diffractograph?

Reference

Walter J. Moore (1967). *Seven Solid States: an introduction to the properties of solids*, A. Benjamin, New York.

2

Matter Waves

In the previous chapter a billiard ball approach was used to investigate symmetrical packings of atoms. There was also a brief introduction to the techniques for analysing crystalline materials. In this chapter we will start to look at the atom in more detail, in order to see how detail at the atomic level is related to the externally observable properties of solids. Throughout the treatment the Rutherford atom will be used; that is, a very small nucleus containing most of the mass which is surrounded by orbiting electrons. The link between the microscopic and macroscopic levels depends on quantum mechanics.

The development of quantum mechanics

This subject has its origins in the absorption and emission of heat in solids. The law of Dulong and Petit states that one atomic weight of an elemental solid (expressed in kg) rises by one degree (independent of temperature) for every 25 kJ of energy absorbed. Classical physics can be used to explain this observation satisfactorily. The law, however, does not hold at low temperatures. When a solid is cooled the specific heat (energy input/temperature rise) decreases and tends towards zero as the absolute zero of temperature is approached.

Classical physics was also unable to account for some of the observations in black body radiation. It predicted that the emission or radiation of heat should continue to rise at shorter wavelengths. This prediction was called the ultraviolet catastrophe paradox. Fortunately, it does not occur.

There were similar problems with an electron orbiting a Rutherford atom. Classical physics predicts that it should spiral into the nucleus, just like a satellite returning to earth. Again, it is perhaps fortunate that this does not occur.

The solution was provided by Max Planck, who suggested a resolution of the specific heat problem. A solid does not consist of a collection of vibrating atoms with a continuous spectrum of frequencies. Rather, the frequencies are fixed and in order to change from one frequency to another an atom has to absorb or emit a discrete package or quantum of vibrational energy. The energy of a quantum is proportional to the frequency of oscillation.

$$E = hf$$

where f = frequency and h is a constant, Planck's constant = 6.62×10^{-34} J s. The quantum of vibrational energy is called a *phonon*.

This was immediately used to solve both the specific heat and black body radiation problems.

Planck's formula was used by Bohr in his model for the hydrogen atom. He proposed a structure based on the following postulates:

1. Electrons move in closed orbits of fixed radius around a nucleus.
2. Electrons may only move from one closed orbit to another by the emission or absorption of a quantum of energy ($E = hf$).
3. The angular momentum (mvr) of an electron in a closed orbit is $nh/2\pi$ (m is the electron mass, v its velocity and r its radius from the nucleus. n is an integer and h is Planck's constant).

This treatment gave an adequate explanation for many of the observations of hydrogen atoms. It gave a value for the radius at different energy levels. The Ritz-Rydberg formula

$$f = \frac{\Delta E}{h} = R \left[\frac{1}{n_2^2} - \frac{1}{n_1^2} \right]$$

$$R = \frac{me^4}{8\epsilon_0^2 \, \epsilon_r^2 \, h^2} \quad \text{(the Rydberg constant)}$$

n_1, n_2 are quantum numbers

can be derived using Bohr theory and accurately predicts the observed spectral lines of hydrogen.

The Bohr atom proved to be of immense value in chemistry. It was now possible to construct an atom with an appropriate number of protons which determined the mass. An equal number of electrons could then be inserted into the Bohr orbits, subject to certain rules. It was found that the arrangement of the outermost electrons determined the observed chemical properties of any particular element. (Lithium, sodium, potassium and rubidium each have one electron in the outermost shell; they have similar chemical properties.)

Armed with these successes, quantum mechanics was about to extend its frontiers.

The duality concept

The dual nature of light had been accepted, although not understood, before the early part of this century. Newton's corpuscular theory could account for reflection, refraction and some aspects of the photo-electric effect. Huygens' wave theory could account for reflection, refraction and diffraction. Planck's hypothesis explained apparent anomalies in the temperature dependence of the specific heats of solids. Einstein extended the concept to the subject of light and showed how the experimental observations of the photo-electric effect could be explained in terms of this hypothesis. Thus light can be considered to be quantised into wave packets, called *photons*, each with energy hf.

de Broglie's hypothesis

Until 1924, there seemed to be no doubt that matter was particulate in nature. However, in that year Prince Louis de Broglie suggested a revolutionary new concept. Matter was essentially wave-like and had a wavelength given by

$$\lambda = h/P$$

where h is Planck's constant and P is the momentum (mv) of the particle.

So far as large particles are concerned, the statement is of little consequence. A 1 kg projectile travelling at 500 m/s has a wavelength of 1.32×10^{-36} m, which is obviously negligible. However, the situation is quite different for small (that is, sub-microscopic particles), especially if they are travelling at high speeds.

Consider a 3.6 kV electron

$$E = 3.6 \text{ keV} = mv^2/2$$
$$P = (2mE)^{1/2} = h/\lambda \text{ (}m\text{ is the electron mass)}$$

Thus $\lambda = 2 \times 10^{-11}$ m = 0.2 Å.

This may appear small, but it is of the same order of magnitude as the inter-atomic spacings in crystals. It is almost identical to the wavelength of X-rays used in crystallography experiments and it is obvious that such a high-energy beam of electrons could be used instead of X-rays.

Confirmation of de Broglie's hypothesis

Experimental confirmations of de Broglie's hypothesis are all based on observing the diffraction properties of 'matter waves'. The first of these was the Davisson-Germer experiment (Davisson and Germer, 1927) which is analogous to the Laue technique of X-ray diffraction. A single crystal of nickel had been bombarded with an electron beam (see figure 2.1). Davisson and Germer were able to use the de Broglie hypothesis to explain their observations. They found that the scattered current depended on the bombarding current, the bombarding potential, the scattering angle θ and the orientation ϕ of the crystal. The incident current dependence was a simple proportionality. The dependence on φ was determined by the symmetry of the crystal structure. Two separate sets of results were obtained for incident beams, a set of three in the (111) direction and a set of three of different intensity in the (100) direction. The (111) polar and azimuthal scattering results are shown in figure 2.2.

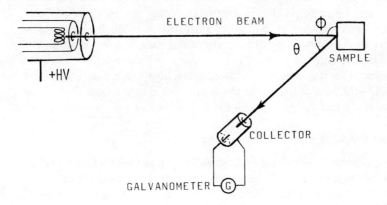

Figure 2.1 Schematic outline of the Davisson–Germer experiment.

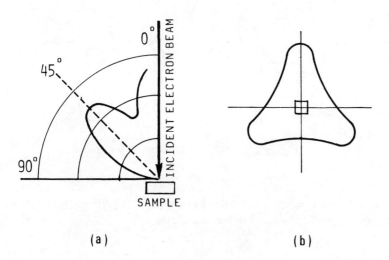

Figure 2.2 (a) Polar scattering of a 54 V electron beam by (111) planes of nickel. (b) Azimuthal scattering of a 54 V electron beam by (111) planes of nickel.

The de Broglie formula was used to calculate a value for λ. It was then possible to show that the scattering angle was identical to the value that would have been obtained from the Bragg formula for X-rays of the same wavelength.

There have been further confirmations of the de Broglie formula. The Thompson experiment was equivalent to the Debye–Scherrer method for X-ray diffraction (Thompson, 1928). Nevertheless, it could be argued that the wave-like nature might be peculiar to electrons or other charged particles. The question still remained as to whether other matter would display the same properties. The wave-like nature of atoms and molecules was investigated by Stern (Estermann and Stern, 1930). In their experiments a fine stream of atoms or molecules was ejected from an oven into a high vacuum. A series of slits was placed in the path of the beam, which gave rise to diffraction. This proved conclusively that the de Broglie formula was not just a property of electrons.

The de Broglie formula and Bohr's third postulate

Bohr's original theory of the hydrogen atom did not have a firm theoretical basis for the third postulate ($mvr = nh/2\pi$). It was empirical in that it gave the correct answers. If, however, the moving electron can be assumed to have wave nature, then the de Broglie formula can be invoked.

An electron wave situated at a fixed radius from a nucleus can be stable only if it is, in fact, a standing wave.

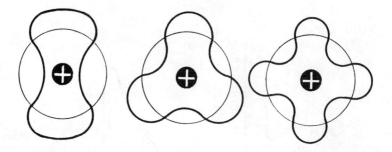

Figure 2.3 Standing waves on a ring.

A condition for a standing wave to exist on a ring is that the circumference must be an integral number of wavelengths; that is

$$2\pi r = n\lambda \quad \text{or} \quad \lambda = 2\pi r/n$$

de Broglie's hypothesis, $mv = h/\lambda$, means that $\lambda = h/mv$, thus

$$2\pi r/n = h/mv$$

Thus, if an electron orbiting a nucleus has wave-like nature, then a condition for

MATTER WAVES

its stability as a standing wave is that

$$mvr = nh/2\pi$$

The resolution of a paradox

The Davisson–Germer and Stern experiments all prove that matter exists as waves. However, Thompson's e/m measurement and Millikan's oil drop experiment prove that an electron is a definite particle with a definite mass and charge. This apparent paradox can be resolved by treating matter as a packet composed of a group of matter waves of varying amplitude and frequency.

The formation of a particular wave from a series of component waves is achieved using a Fourier synthesis technique similar to that used in crystallographic analysis.

Fourier's theorem

Periodic or single pulses of any shape other than sinusoidal form may be represented by a series of sine and cosine waves of varying amplitude whose frequencies are simple ratios of a fundamental frequency. That is, any given function $f(x)$ that is repetitive between $-\pi$ and $+\pi$ can be expressed as

$$f(x) = \frac{a_0}{2} + \sum_{}^{\infty} (a_n \cos nx + b_n \sin nx)$$

$$a_0 = \frac{1}{\pi} \int_{-\pi}^{\pi} f(x)\, dx$$

$$a_n = \frac{1}{\pi} \int_{-\pi}^{\pi} f(x) \cos nx\, dx$$

$$b_n = \frac{1}{\pi} \int_{-\pi}^{\pi} f(x) \sin nx\, dx$$

Thus a saw-tooth waveform where $f(x) = bx$ in the range $-\pi$ to $+\pi$ has

$$f(x) = 2b \left[\sin x - \frac{\sin 2x}{2} + \frac{\sin 3x}{3} - \ldots \right]$$

Note that to define a continuous wave form only requires the specification of components between $-\pi$ and $+\pi$.

In communications, signal processing, acoustics and many other areas it is often convenient to display the Fourier components as an amplitude–frequency spectrum. Figure 2.5 shows such a spectrum for the saw-tooth wave of figure 2.4.

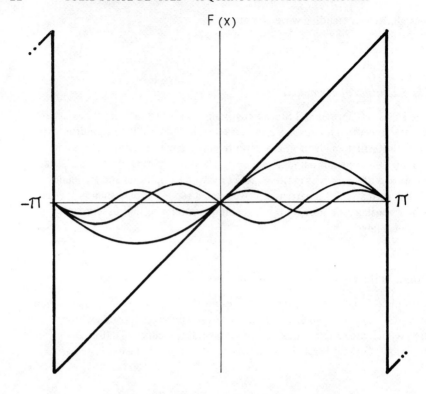

Figure 2.4 A saw-tooth waveform with some of its sine components.

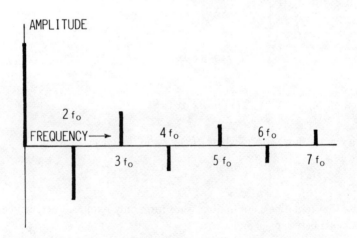

Figure 2.5 Amplitude-frequency spectrum showing the components of the saw-tooth waveform of figure 2.4.

The situation becomes more complex for localised waveforms. In this case, all components between $-\infty$ and $+\infty$ need to be specified. This requires the use of Fourier's integral theorem

$$f(x) = \frac{1}{2\pi} \int_{-\infty}^{\infty} f(k) \exp(jkx) \, dk$$

where k is the wave number $= 2\pi/\lambda$, so that $\exp(jkx)$ represents the motion of a wave in space as will be shown later.

As a consequence of Fourier's integral theorem, any spatial distribution of waves, $f(x)$, has a corresponding distribution of frequencies, $f(k)$, which can be represented in an amplitude-frequency spectrum. The distribution, $f(k)$, in frequency space can be calculated from a spatial distribution $f(x)$ using the inverse Fourier integral theorem

$$f(k) = \frac{1}{2\pi} \int_{-\infty}^{\infty} f(x) \exp(-jkx) \, dx$$

As was mentioned previously, Fourier methods are used extensively in crystallographic analysis. Since the wave number, k, has units of reciprocal length, it should be obvious why it is convenient to use reciprocal lattices. In this work, reciprocal lattice space will normally be called k-space. A distribution in k-space is identical to an amplitude-frequency spectrum such as that shown in figure 2.5.

As a specific demonstration of these concepts, Fourier's inverse integral theorem is now used to calculate the amplitude-frequency spectrum corresponding to the Gaussian distribution shown in figure 2.6. The Gaussian or normal distribution is well known in statistics as a means of expressing the spread of a continuous or near-continuous function (characterised by the

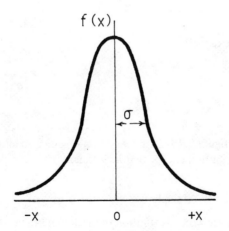

Figure 2.6 A Gaussian or normal distribution.

standard deviation, σ) about a mean. Examples include the distribution of male heights in a large population. This expression could be taken as an approximation of a localised distribution of matter (such as an electron particle) in space.

$$f(x) = A \exp(-x^2/2\sigma^2)$$

$$f(k) = \int_{-\infty}^{\infty} A'\exp(-x^2/2\sigma^2) \exp(-jkx)\, dx$$

$$= A' \int_{-\infty}^{\infty} \exp\left[-\left(\frac{x^2}{2\sigma^2} + jkx\right)\right] dx$$

The constant A' absorbs amplitude, $1/2\pi$ etc. A dummy variable can be inserted into the expression.

$$f(k) = A'\exp\left(-\frac{\sigma^2 k^2}{2}\right) \int_{-\infty}^{\infty} \exp\left[-\left(\frac{x^2}{2\sigma^2} + jkx - \frac{\sigma^2 k^2}{2}\right)\right] dx$$

$$= A'\exp\left(-\frac{\sigma^2 k^2}{2}\right) \int_{-\infty}^{\infty} \exp\left[-\left(\frac{x}{\sqrt{2\sigma}} + \frac{j\sigma k}{\sqrt{2}}\right)^2\right] dx$$

Put

$$M = \left(\frac{x}{\sqrt{2\sigma}} + \frac{j\sigma k}{\sqrt{2}}\right) \text{ so that } dM = \frac{dx}{\sqrt{2\sigma}}$$

Then

$$f(k) = A''\exp\left(-\frac{\sigma^2 k^2}{2}\right) \int_{-\infty}^{\infty} \exp(-M^2)\, dM$$

where $A'' = \sqrt{2\sigma} A'$

$$\int_{-\infty}^{\infty} \exp(-M^2)\, dM = \sqrt{\pi}$$

Therefore

$$f(k) = A'''\exp\left(\frac{-\sigma^2 k^2}{2}\right) \text{ where } A''' = A'' \sqrt{\pi}$$

It can be seen from figure 2.7 that this function is similar in all respects to the original normal distribution except that its standard deviation is $1/\sigma$.

Heisenberg's Uncertainty Principle

The relationship between the distribution of matter and the distribution of its wave components forms the basis of Heisenberg's Uncertainty Principle, which will now be demonstrated for the case of the Gaussian electron.

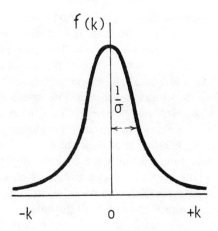

Figure 2.7 Distribution of component frequencies of a Gaussian distribution in k-space.

Using de Broglie's formula, $P = h/\lambda$, it can be shown that

wave number, $k = 2\pi/\lambda = 2\pi P/h = P/\hbar$

where $h/2\pi = \hbar$.

If one takes the example of the normal distribution, the standard deviation in $f(x)$ is, in fact, the uncertainty, Δx, in defining the position of the electron. If the standard deviation of the component matter wave distribution, $f(k)$, is $1/\sigma$, then it can be seen that, since $k = P/\hbar$

$$k = \Delta P/\hbar$$

so that

$$1/\sigma = \Delta k = \Delta P/\hbar = 1/\Delta x$$

or

$$\Delta P \, \Delta x = \hbar/2\pi$$

As the uncertainty in locating the particle is improved, the uncertainty in its momentum increases.

A slightly more rigorous treatment of Heisenberg's Uncertainty Principle gives a more general result

$$\Delta P \, \Delta x \geqslant h/2\pi$$

Heisenberg's Uncertainty Principle becomes an important consideration in any discipline which involves extremely small dimensions. An optical microscope is used to examine specimens greater than about 5 μm. Once the size of the sample is comparable with the wavelength of light used to examine it, diffraction becomes a problem. Nowadays, this problem could be overcome by using a computer-assisted reconstruction of the image from diffraction data. It is usual, however, to use radiation of smaller wavelengths. This is achieved in an electron microscope, where the wavelength is a function of the accelerating voltage. However, as the wavelength is reduced the energy increases, so that a point is eventually reached where any interaction between the specimen and the excitation beam will give rise to a large energy transfer. Thus the excitation required to locate a particle in space gives rise to a momentum change which makes it impossible to give an accurate estimate of position.

In microcircuit manufacture, photolithography is the step where a circuit pattern is defined in a photoresist covering on a silicon slice. Visible light is used for patterns down to about 5 μm. Below that size, electron beam lithography must be used. There are two factors which limit the linewidth resolution and thereby the ultimate scale of miniaturisation of such circuits. The molecules of the polymer which make up the resist have a finite size. Pattern definition smaller than this is impossible. However, well before this limit is reached Heisenberg's Uncertainty Principle has intervened. The electron beam spot has an approximate diameter but, as the energy is increased in order to go to smaller wavelengths, the uncertainty about spot size increases and it is this which ultimately limits the ability to define ever smaller linewidths.

Problems

2.1. Find the de Broglie wavelength of the following particles:
    (i) An electron in a semiconductor having average thermal energy at $T = 300$ K of $\frac{1}{2}kT$ and possessing an effective mass of $\frac{1}{2}m_e$ (rest mass of electron).
    (ii) A neutron having thermal energy of $\frac{1}{2}kT$ at $T = 300$ K (neutron mass = 1.67×10^{-27} kg = 1 a.m.u.).
    (iii) An α-particle (He_2^4 nucleus) of kinetic energy 10 MeV (10^7 eV) (α-particle mass = 4 a.m.u.).

2.2. The thermal neutrons of problem 2.1(i) are directed towards the d_{111} planes of a crystal of aluminium (see problem 1.1). Calculate the scattering angle for first-order diffraction.

2.3. Problem 1.5 describes a measurement where 1.54 Å X-rays were used. Repeat the exercise for the case where the X-rays are replaced by a 50 kV electron beam.

2.4. de Broglie's hypothesis suggests that the higher the energy of a particle, the shorter the wavelength associated with it.

MATTER WAVES

An optical microscope is limited in resolution by the wavelength (λ) of light which is used. It cannot resolve matter whose dimensions are of the same order or smaller than λ. Electron microscopes are capable of considerably better resolution.

What is the smallest dimension that can normally be resolved using 50 kV electrons?

2.5. 10 MeV α-particles are used in a scattering experiment. Calculate the minimum dimension that can be resolved.

2.6. In a Stern diffraction experiment, a beam of nickel atoms (atomic mass = 58.7) having energy $\frac{1}{2}kT$ at 3000 K is directed past a series of slits. Assuming that the Bragg diffraction formula can be applied, estimate the separation between the slits if the angle for first-order diffraction is 30 seconds of arc.

2.7. Derive Fourier components for $y = x^2$ which is continuous between $\pm\pi$.

2.8. Sketch the spatial distribution of $f(x) = \pi x^2 \exp(-x^2/2)$ and calculate the corresponding distribution in frequency space. *Hint:*

$$\int_{-\infty}^{\infty} x^2 \exp(-ax^2)\,dx = \frac{1}{2a}\sqrt{\frac{\pi}{a}} \quad \text{and} \quad \int_{-\infty}^{\infty} x \exp(-x^2)\,dx = 0$$

2.9. A 10 MeV α-particle is emitted within 10^{-12} s. How many wavelengths are there in a wave packet? *Hint:* how many wavelengths appear within 10^{-12} s?

2.10. A 1 MeV γ-ray (electromagnetic radiation $E = 10^6$ eV) is emitted within a time uncertainty of 10^{-12} s. Calculate the length of a wave packet.

References

C. J. Davisson and L. H. Germer (1927). *Phys. Rev.*, **30**, 705–740.
I. Estermann and O. Stern (1930). *Z. Phys.*, **61**, 95–125.
G. P. Thompson (1928). *Proc. Roy. Soc.*, **117A**, 600.

3

Wave Mechanics

Introduction

The material of the previous chapter has set out the foundations of quantum mechanics. It has been shown that a particle has wave character. Once this is accepted, it becomes possible to use Fourier synthesis to construct a hypothetical particle. Since matter has been shown to have wave-like properties and since it is possible to define it by means of the superposition of component waves, it follows that it should be possible to use the mathematical tools which are available for treating wave motion. This forms the basis of wave mechanics.

There are a variety of approaches to this subject. One can present the Schrödinger wave equation as a definition and build the theory from there. This approach can be informative but is somewhat restricted. Alternatively, one can use a more rigorous treatment in which all macro-measurables are presented as statistical functions of what is happening at the microscopic level. The approach which is used here attempts a middle course (in terms of complexity) which should nevertheless demonstrate that what is presented is in fact only a part of a much broader subject.

The first part of this chapter introduces the various techniques for manipulating waves. Wave functions and operators are presented and the Schrödinger equation is derived. The remainder of the chapter is concerned with its solution for a number of simple, but nevertheless important cases.

Review of wave motion

Before going into too much detail it may first be useful to review some of the salient features of sinusoidal wave motion.

A wave $Y = A \sin \theta$ can be expressed in terms of time by putting $\theta = \omega t$. The motion can also be expressed as a function of distance, x. The angular frequency,

WAVE MECHANICS

ω, is related to wavelength, λ, and wave velocity, v, by the formula

$$f = v/\lambda \qquad \omega = 2\pi f \qquad f = \text{frequency}$$

therefore

$$Y = A \sin(2\pi v t/\lambda)$$

Provided there is no acceleration, $v = x/t$. Therefore

$$Y = A \sin(2\pi x/\lambda)$$

$2\pi/\lambda$ has units of reciprocal length. It is called the *wave number* k. If λ is in metres, k is the number of waves in 2π metres. It is a very convenient measure and is used extensively. Hence

$$Y = A \sin(\omega t) \text{ is equivalent to } A \sin(kx)$$

The motion of a wave along a vibrating string is a well-known mechanical example which provides a useful analogue. The time and position dependent motion of any point, $Y(x, t)$, can be expressed by a second-order partial differential equation

$$\frac{\partial^2 Y(x, t)}{\partial t^2} = v^2 \frac{\partial^2 Y(x, t)}{\partial x^2}$$

If $Y(x, t) = \phi(x) \exp(j\omega t)$, then

$$\frac{d^2 \phi}{dx^2} + \left(\frac{\omega}{v}\right)^2 \phi = 0$$

which is a time-independent form of the wave motion

If $\omega = 2\pi f$ and $v = f\lambda$, then

$$\frac{d^2 \phi}{dx^2} + \left(\frac{2\pi}{\lambda}\right)^2 \phi = 0$$

$\phi(x)$ is often called an *eigenfunction* — that is, a single-valued function which satisfies the wave equation.

If $\phi(x) = A \sin(2\pi x/\lambda) + B \cos(2\pi x/\lambda)$ is substituted into the equation, it will be clear that it provides a satisfactory general solution. The solution can be made more specific by the use of boundary conditions in the following way.

If the wave is restricted so that it can only exist on the string of length, L, then at $x = 0$, $\phi = 0$. Since $\cos(0) = 1$, B must be zero. At $x = L$, $\phi = 0$. This can only happen if there is no wave (a perfectly satisfactory solution) or if $2\pi L/\lambda = n\pi$, where n can assume integral values.

The solution of the equation for wave motion on a string is therefore

$$\phi_n = A \sin \frac{n\pi x}{L} \qquad (n = 0, 1, 2 \ldots)$$

The example shows that there are restrictions placed on the possible values of ϕ. For each eigenfunction there is a corresponding $\lambda = 2L/n$, which is often called an *eigenvalue*.

Wave functions

Since a particle is a wave, it can be described by a wave function. ψ (psi) is the symbol used for the eigenfunction, which defines our total knowledge of the wave. It can either be considered to represent a particle wave or the result of a superposition of component waves.

They can be moderately simple functions which can be real, imaginary or complex. The following are some examples

$\psi = A \cos(kx - \omega t)$ is a wave function

$\psi = A \cos(kx - \omega t) + B \sin(kx - \omega t)$ describes a plane wave moving in the x direction

$\psi = A \cos(kx + \omega t) + B \sin(kx + \omega t)$ describes a plane wave moving in the $-x$ direction

In order to describe a particle moving in space, one must use a complex wave function.

It can be shown from the examples above that

$\psi = A \exp[j(kx - \omega t)]$ describes a wave (particle) moving in the x direction

$\psi = B \exp[-j(kx + \omega t)]$ describes a wave (particle) moving in the $-x$ direction

For simplicity it will be assumed that the wave functions for moving particles are at all times identical to the wave function at $t = 0$. Thus the wave functions for particle motion in the x and $-x$ directions can be written as $A \exp(jkx)$ and $B \exp(-jkx)$ respectively.

As the actual expressions for wave functions can sometimes be extremely complicated (see figure 3.8), it is usual to do all mathematical manipulations using ψ, and to insert the proper expression only when absolutely necessary.

Born was the first to suggest that there was a probability associated with a wave function. ψ itself cannot be a probability. By definition, a probability must be real and positive. Wave functions may be real, imaginary or complex. It is therefore necessary to define a conjugate ψ^* such that the product $\psi\psi^*$ is always real.

Having shown that the Davisson–Germer and Thompson experiments confirm that a collection of high-energy electrons which impinge on a screen can be treated using the de Broglie formula, one can then interpret the resulting intensity pattern as a statistical sum of probabilities of locating the individual electron waves at any position on the screen. In electromagnetics, the intensity of radiation is proportional to the square of the wave amplitude. By analogy $\psi\psi^*$ is interpreted as the probability of locating the particle in a volume of space.

As a first example, one could consider the function $\psi = A \sin \omega x$. $A^2 \sin^2 \omega x$ therefore represents a probability function. From figure 3.1 it is clear that the probability function is a minimum when $d\psi/dx$ is a maximum (shadows of Heisenberg: the larger the momentum, the smaller is the probability of being able to define position).

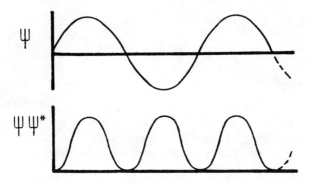

Figure 3.1   Plots of ψ and $\psi\psi^*$ versus x for $\psi = A \sin kx$.

$\psi = A \sin \omega x$ in the previous example is known as an unnormalised wave function. A wave function has its existence within a defined region. Once the boundaries of that region are known then the probability, given by $\psi\psi^*$, must be unity inside those limits.

As a second example consider a particle described by the wave function

$$\begin{aligned}\psi &= A \exp(-x^2/2\sigma^2 + jkx) \quad \sigma = \text{constant} \\ \psi^* &= A \exp(-x^2/2\sigma^2 - jkx) \\ \psi\psi^* &= A^2 \exp(-x^2/\sigma^2)\end{aligned}$$

A normalised wave function has the property

$$\int_{\substack{\text{all}\\\text{space}}} \psi\psi^* = 1$$

so

$$A^2 \int_{-\infty}^{\infty} \exp(-x^2/\sigma^2)\,dx = A^2 \sigma \sqrt{\pi} = 1$$

and therefore

$$A = \frac{1}{\sqrt{\sigma\sqrt{\pi}}}$$

A is a normalising constant. The normalised wave function is now

$$\psi = \frac{1}{\sqrt{\sigma \sqrt{\pi}}} \exp[-x^2/2\sigma^2 + jkx]$$

In concluding this section one can summarise three requirements for suitable wave functions:

(1) They must be single-valued functions — that is, $\psi\psi^*$ can only have one value.
(2) They must be continuous.
(3) The integral $\int \psi\psi^*$ must be finite.

Observables and operators

Observables are defined as the normal mechanical properties which can be measured, such as position, momentum, energy etc. In wave mechanics observables are represented by operators which are merely mathematical instructions. The application of operators can be summarised by the following three postulates.

(1) To every observable there corresponds an operator — see table below.

Observable	Classical representation	Quantum mechanical representation	Measured value
Position	x, y, z etc.	$[X] = x$ etc.	x, y, z etc.
Momentum	$P_x = mv_x = m\dfrac{dx}{dt}$	$[P_x] = \dfrac{h}{2\pi j}\dfrac{\partial}{\partial x}$	P_x etc.
Total energy $E = KE + PE$	$H = KE + PE = \dfrac{p^2}{2m} + V$ $= \dfrac{1}{2m}(P_x^2 + P_y^2 + P_z^2) + V$	$[H] = \dfrac{-h^2}{8\pi^2 m}\nabla^2 + V$ $\nabla^2 = \dfrac{\partial^2}{\partial x^2} + \dfrac{\partial^2}{\partial y^2} + \dfrac{\partial^2}{\partial z^2}$	E

(2) The only possible values measurable for an observable are those for which

[operator] ψ = (measured value) ψ

The measured value is the eigenvalue corresponding to the eigenfunction, ψ.

(3) The expected mean value of a sequence of many measurements of an observable having an operator $[Q]$ is

$$\bar{Q} = \frac{\int_{\text{space}} \psi^*[Q]\psi \, d\tau}{\int_{\text{space}} \psi^*\psi \, d\tau} \quad (d\tau \text{ is the element of volume})$$

WAVE MECHANICS

If ψ is normalised then the denominator is unity. Thus, if the wave function is known, one can predict the average of many measurements of an observable. One could therefore predict the intensity pattern that might be expected in a Davisson–Germer, Thompson or Stern experiment.

One can now consider some examples of the second postulate.

Position operator

If the measured value of position in the x direction is x_k then

$$[X]\psi = x\psi = x_k\psi$$

and therefore

$$(x - x_k)\psi = 0$$

Thus there are no restrictions on the values which x can take.

Momentum operator

(i) In the x-direction

If $P(x, k)$ is the measured momentum in the x direction then

$$[P]\psi = \frac{h}{2\pi j} \frac{\partial}{\partial x} = P(x, k)\psi$$

thus

$$\psi = A \exp\left[\frac{2\pi j P(x_1 k)x}{h}\right]$$

There are no restrictions on the values of allowed momentum.

(ii) Angular momentum

Just as $[P_x] = \dfrac{h}{2\pi j}\dfrac{\partial}{\partial x}$, $[P_\theta] = \dfrac{h}{2\pi j}\dfrac{\partial}{\partial \theta}$

therefore

$$\frac{h}{2\pi j} \frac{\partial \psi}{\partial \theta} = P_{(\theta, k)}\psi$$

so that

$$\psi = A \exp\left[\frac{2\pi j}{h} P_{(\theta, k)}\psi\right]$$

In order that ψ be a single-valued function, ψ at $\theta = 0$ must be the same as ψ at $\theta = n\pi$, $n = $ integer. That is

$$P_{(\theta, k)} = \frac{nh}{2\pi} \quad (n = 0, 1, 2 \ldots \psi)$$

which is identical to Bohr's third postulate.

The definition of $[P]$ can be extended into three dimensions

$$[P] = [P_x] + [P_y] + [P_z]$$

$$= \frac{h}{2\pi j} \left[\frac{\partial}{\partial x} + \frac{\partial}{\partial y} + \frac{\partial}{\partial z} \right]$$

The total energy or Hamiltonian operator

The total energy of a system is the sum of the kinetic and potential energies

$$E = \tfrac{1}{2}mv^2 + V \quad (V \text{ is the potential energy})$$

$$= \frac{P^2}{2m} + V \quad (P \text{ is the momentum})$$

The total energy operator, $[H]$, which is often called the Hamiltonian operator after the 19th century Irish mathematician and astronomer, Sir William Rowan Hamilton, has been defined as

$$[H] = \left[\frac{-\hbar^2}{2m} \nabla^2 + V \right]$$

where

$$\nabla^2 = \frac{\partial^2}{\partial x^2} + \frac{\partial^2}{\partial y^2} + \frac{\partial^2}{\partial z^2}, \quad \hbar = \frac{h}{2\pi}$$

Thus $[H]\psi = E\psi$ can be rewritten as

$$\left[-\frac{\hbar^2}{2m} \nabla^2 + V \right] \psi = E\psi$$

$$-\frac{\hbar^2}{2m} \nabla^2 \psi + (V - E)\psi = 0$$

$$\nabla^2 \psi + \frac{2m}{\hbar^2} (E - V)\psi = 0$$

The last expression which contains the total energy E and the potential energy V is called Schrödinger's wave equation. It is a second-order differential equation which, if it can be solved, can give E or ψ.

In one dimension the wave equation is

$$\frac{d^2 \psi}{dx^2} + \frac{2m}{\hbar^2} (E - V)\psi = 0$$

It is not much different from the second-order differential equation for the motion of a wave on a string and it should be possible to solve it in a similar way.

Wave mechanics

Wave mechanics is concerned with solving the Schrödinger wave equation. This is not particularly difficult in essence, although when applied to sufficiently complex situations it may be almost impossible to solve by analytical methods. Before starting wave mechanics it is useful to bear some things in mind.

Many of the examples which will be considered in this section have analogies in other disciplines. A wave-like particle travelling in space is not dissimilar to the propagation of electromagnetic waves. If the particle experiences a change in potential it may undergo reflection. If microwaves in a waveguide experience an impedance transition, they too will be reflected. Similarly, light is reflected from a surface where there is a change in refractive index.

It will be seen that the absorption of a particle within a potential barrier is not much different from the situation where light encounters an absorbing medium.

The intensity of light I_t, which is transmitted through length l of a medium of extinction coefficient k is given by the Beer-Lambert law

$$I_t = I_0 \exp(-kl)$$

I_0 is the intensity of the incident light

The absorption of light by a solution of ink is a well-known example of the manifestation of this law. The amount of light which is transmitted depends on the incident intensity, and the length of the absorbing medium. The extinction coefficient, k, depends on the amount of ink in solution and affects different wavelengths differently. Thus blue ink transmits blue light in preference to other wavelengths.

The second thing that should be borne in mind is that the wave equation does not always require an exhaustive solution. It has already been explained that a wave function ψ gives the total knowledge of the system. The majority of this information can be obtained during a solution. However, if one only wishes to estimate the energy of a particle under given conditions, a total derivation of the equation may be a tedious and superfluous exercise.

During solution the wave equation is usually written in one dimension as

$$\frac{d^2\psi}{dx^2} + k^2\psi = 0 \text{ where } k \text{ is the wave number}$$

$$k = \left[\frac{2m(E-V)}{\hbar^2}\right]^{1/2}$$

This can be proved as follows.

Kinetic energy = total energy − potential energy

$$\tfrac{1}{2}mv^2 = E - V = \frac{P^2}{2m}$$

$$\frac{2m(E-V)}{\hbar^2} = \frac{2m}{\hbar^2} \times \frac{P^2}{2m} = \frac{P^2}{\hbar^2}$$

Using de Broglie's hypothesis it can be shown that

$$\frac{P^2}{\hbar^2} = [2\pi/\lambda]^2 = k^2$$

Boundary conditions for solution of Schrödinger's equation

The wave equation expresses the behaviour of the particle throughout all space. The method used in solution is to put in arbitrary solutions (remember that $y = A \sin \omega t$ is an arbitrary solution of the equation for simple harmonic motion). Any prior information we may have about the motion can be used to help turn this into an exact solution.

As an example, if it is known that an electron is constrained to exist only inside a cube of length a, one can immediately say that $\psi = 0$, except between 0 and a, and one can also say that the momentum of the electron must be zero except between 0 and a, otherwise it would transfer momentum to its surroundings, and either escape or tend towards zero energy. From the definition of the momentum operator, if $P(x) = 0$, $d\psi/dx = 0$. The knowledge that $\psi = 0$ except between 0 and a and $d\psi/dx = 0$ between the same limits gives boundary conditions which we impose from our knowledge (intuitive or otherwise) of the physical system, and which are useful in obtaining an exact solution.

In the following pages, several specific examples are considered. Although these have been chosen to demonstrate the techniques of wave mechanics, they provide results which will be utilised later in this text. For simplicity, all solutions will be in one dimension only.

Particle encountering a potential step

Case 1: E (energy of particle) $> V_1$ *and* $E > V_2$

Region 1 covers all space where the particle is influenced by potential V_1 — that is, to the left of $x = 0$. A change to potential V_2 occurs at this point and covers all space in region 2.

The wave equation in region 1 where wave number is k_1 is

$$\frac{d^2 \psi}{dx^2} + \frac{2m}{\hbar^2}(E - V_1)\psi = 0$$

or

$$\frac{d^2 \psi}{dx^2} + k_1^2 \psi = 0$$

An equation of this form has the following solutions

WAVE MECHANICS

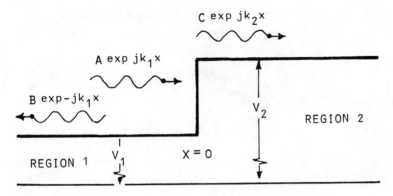

Figure 3.2 A potential step.

$\psi_1 = A \exp(jk_1 x)$ — a wave travelling in the x direction

$\psi_1 = B \exp(-jk_1 x)$ — a wave travelling in the $-x$ direction

The most general solution is

$$A \exp(jk_1 x) + B \exp(-jk_1 x)$$

The wave equation in region 2 where the wave number is k_2 is

$$\frac{d^2 \psi}{dx^2} + k_2^2 \psi = 0$$

$$k_2^2 = \frac{2m}{\hbar^2}(E - V_2)$$

This has similar solutions of which the most general is

$$\psi_2 = C \exp(jk_2 x) + D \exp(-jk_2 x)$$

Our intuitive knowledge indicates that a wave incident from the left can be either reflected or transmitted. The transmitted wave is travelling in the x direction so that only $\psi_2 = C \exp(jk_2 x)$ has physical meaning.

Detailed information can be determined using boundary conditions as follows.

1. At $x = 0$, $\psi_1 = \psi_2$ matter is conserved.
The wave does not suddenly change at $x = 0$. Therefore

$$A + B = C \tag{3.1}$$

2. At $x = 0$, $\dfrac{d\psi_1}{dx} = \dfrac{d\psi_2}{dx}$ momentum is conserved.

The momentum of the wave does not suddenly change at $x = 0$. Therefore

$$k_1 A - k_1 B = k_2 C \tag{3.2}$$

Equations (3.1) and (3.2) can be used to eliminate B and C to give the reflected and transmitted waves in terms of the amplitude of the incident wave.

$$B = \left[\frac{k_1 - k_2}{k_1 + k_2}\right] A \qquad C = \left[\frac{2k_1}{k_1 + k_2}\right] A$$

Incident wave $= A \exp(jk_1 x)$

Reflected wave $= \left[\dfrac{k_1 - k_2}{k_1 + k_2}\right] A \exp(-jk_1 x)$

Transmitted wave $= \left[\dfrac{2k_1}{k_1 + k_2}\right] A \exp(jk_2 x)$

This result is very similar to the situation in electromagnetic wave theory when a wave experiences a transition in impedance from z_1 to z_2. In this case the reflection coefficient is

$$P = \frac{z_1 - z_2}{z_1 + z_2}$$

Case 2: E (energy of particle) $> V_1$ but $E < V_2$
The solution in region 1 is exactly the same as in case 1.
The solution in region 2 is initially as in case 1, namely

$$\psi_2 = C \exp(jk_2 x) + D \exp(-jk_2 x)$$

Again, intuitive knowledge is implemented. Experience suggests that if a particle encounters a barrier which it cannot surmount, it does not get past it. Put as a boundary condition in its extreme, it can be said that

$$\text{as } x \to \infty \qquad \psi_2 \to 0$$

This can only be true if $C = 0$, so that the solution becomes

$$\psi_2 = D \exp(-jk_2 x)$$

and   since $E < V_2$

$$k_2^2 = \frac{2m}{\hbar^2}(E - V_2) < 0$$

This means that

$$k_2^2 = \frac{4\pi^2}{\lambda_2^2} < 0$$

λ_2 is the particle wavelength in region 2 which must be imaginary.
A modulus of wavelength can be defined so that

$$\lambda_2 = j |\lambda_2|$$

therefore

$$k_2 = \frac{2\pi}{j |\lambda_2|}$$

This can be inserted into $\psi_2 = D \exp(-jk_2 x)$, and then becomes

$$\psi_2 = D \exp\left(-\frac{2\pi x}{|\lambda_2|}\right)$$

which represents a decay function, indicating that the wave is attenuated inside the barrier (figure 3.3). Its depth of penetration depends on the value of λ_2.

The analogy with the Beer–Lambert law should now be clearer. There is also an analogy with the electromagnetic phenomenon of skin effect. The flow of current in a conductor at high frequencies is limited to a small region close to the surface. In fact, the current decays exponentially from the surface. The decay constant, δ, depends on the frequency as well as the electric and magnetic properties of the conductor.

We have now demonstrated that the amplitude of a wave decays to zero inside a potential barrier of infinite length. Accordingly, if a particle is incident on a barrier of non-infinite thickness, there should be some probability that it will have non-zero amplitude when it arrives at the other side of the barrier. This is the basis of quantum mechanical tunnelling which will be considered next.

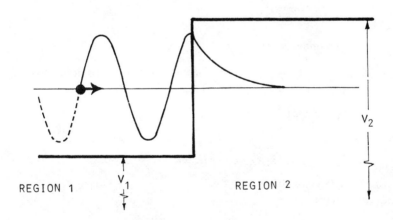

Figure 3.3 Penetration of a potential barrier.

Quantum mechanical tunnelling through a finite barrier

The regions which are considered are shown in figure 3.4. Regions 1 and 3, being free space both have the same wave number, taken as k_1.

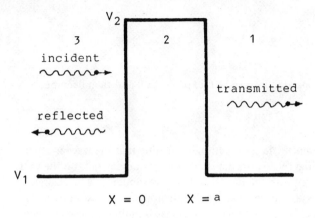

Figure 3.4 A finite thickness potential barrier.

Solution in region 1

$$\psi_1 = A \exp(jk_1 x)$$

In region 2, $E < V_2$ so that

$$\frac{2m}{\hbar^2}(E - V_2) < 0$$

Thus the wave number is imaginary.

The algebraic manipulation is made somewhat easier if k_2' is defined as the imaginary component $= jk_2$, where k_2 is real.

$$\psi_2 = B \exp(jk_2' x) + C \exp(-jk_2' x)$$

Therefore

$$\psi_2 = B \exp(-k_2 x) + C \exp(k_2 x)$$

Using boundary conditions at $x = a$
(i) $\psi_2(a) = \psi_1(a)$

$$B \exp(-k_2 a) + C \exp(k_2 a) = A \exp(jk_1 a) \tag{3.3}$$

(ii) $\dfrac{d\psi_2(a)}{dx} = \dfrac{d\psi_1(a)}{dx}$

$$B \exp(-k_2 a) - C \exp(k_2 a) = -j \frac{k_1}{k_2} A \exp(jk_1 a) \tag{3.4}$$

WAVE MECHANICS

Solving in terms of A

$$B = \frac{A}{2}\left(1 - j\frac{k_1}{k_2}\right) \exp[(jk_1 + k_2)a]$$

$$C = \frac{A}{2}\left(1 + j\frac{k_1}{k_2}\right) \exp[(jk_1 - k_2)a]$$

In region 3

$$\psi_3 = D\exp(jk_1 x) + E\exp(-jk_1 x)$$

(k_1 because regions 1 and 3 consist of the same medium.)
Using boundary conditions at $x = 0$

(i) $\psi_3(0) = \psi_2(0)$, therefore

$$D + E = C + B \tag{3.5}$$

(ii) $\dfrac{d\psi_3(0)}{dx} = \dfrac{d\psi_2(0)}{dx}$, therefore

$$D - E = -j\frac{k_2}{k_1}(C - B) \tag{3.6}$$

Solving for D and E in terms of B and C gives

$$D = \frac{B}{2}(1 + jk_2/k_1) + \frac{C}{2}(1 - jk_2/k_1) \tag{3.7}$$

$$E = \frac{B}{2}(1 - jk_2/k_1) + \frac{C}{2}(1 + jk_2/k_1) \tag{3.8}$$

Replacing B and C by the relevant terms in A gives

$$D = \frac{A}{4}\left[\left(1 - j\frac{k_1}{k_2}\right)\left(1 + j\frac{k_2}{k_1}\right)\exp(jk_1 + k_2)a + \left(1 - j\frac{k_2}{k_1}\right)\left(1 + j\frac{k_1}{k_2}\right)\exp(jk_1 - k_2)a\right]$$

$$= \frac{A\exp(jk_1 a)}{4}\left[\left(1 - j\frac{k_1}{k_2}\right)\left(1 + j\frac{k_2}{k_1}\right)\exp(k_2 a) + \left(1 - j\frac{k_2}{k_1}\right)\right.$$

$$\left.\left(1 + j\frac{k_1}{k_2}\right)\exp(-k_2 a)\right]$$

If $k_2 a$ is large, then the second term in brackets may be neglected.

$$DD^* = \frac{AA^*}{16}\left[\left(1 + \frac{k_1^2}{k_2^2}\right)\left(1 + \frac{k_2^2}{k_1^2}\right)\exp(2k_2 a)\right]$$

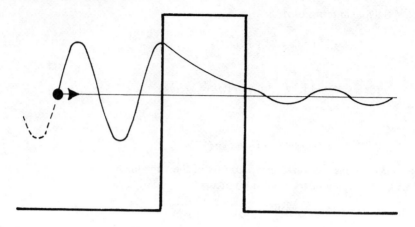

Figure 3.5 Quantum mechanical tunnelling through a barrier. Note that the multiple internal reflections which occur at $x = 0$ and $x = a$ have been omitted for the sake of clarity.

The transmission coefficient, defined as the ratio of the square of the amplitudes of transmitted and incident waves, is effectively the quantum mechanical tunnelling probability. It is given by

$$\frac{AA^*}{DD^*} = \frac{16 \exp(-2k_2 a)}{\left[1 + \frac{k_1^2}{k_2^2}\right]\left[1 + \frac{k_2^2}{k_1^2}\right]}$$

Quantum mechanical tunnelling manifests itself in many branches of physics and electronics.

One can postulate the existence of a barrier between the nucleus of an atom and the outside world. Under most circumstances this inhibits the escape of components of the nucleus. The emission of an α-particle then occurs as a result of tunnelling through the barrier.

One particular semiconductor device, the Esaki diode, provides a negative resistance as a result of quantum mechanical tunnelling. Within the solid state part of this text we will see the application of tunnelling in field emission, the Zener diode, the formation of an ohmic contact etc. Indeed, it will be seen that the entire band theory of solids depends on tunnelling.

The next example of the application of wave mechanics is also significant in many areas of physics, chemistry and electronics, and will be important in our development of the band theory of solids.

Particle in a one-dimensional potential well

It will be noted that the previous solutions were terminated once useful information had been derived. In this example the wave equation will be solved more extensively. A particle such as an electron is constrained to exist inside a potential well of length a (see figure 3.6). The potential inside the well $(V_i) = 0$. V_0 is the potential outside the well. Since we have defined the region of existence of the particle, we can say that $\psi = 0$ outside the well. Schrödinger's wave equation is obeyed inside the well.

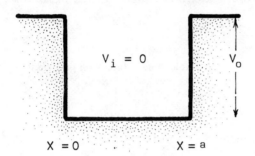

Figure 3.6 A one-dimensional potential well of length a.

One could use solutions of the form $\exp(\pm jkx)$, but these tend to be retained for particles which are moving. For particles which are constrained as this is, it is more convenient to use sin and cos forms of the general solution.

$$\psi = A \sin kx + B \cos kx \quad \text{where } k^2 = \frac{2mE}{\hbar^2}$$

Boundary conditions are now applied. $\psi = 0$ at $x = 0$, so that B must be zero since $\cos 0 = 1$. The solution then reduces to $\psi = A \sin kx$.
$\psi = 0$ at $x = a$. This condition can only be true if either

1. $A = 0$. This is trivial since it means that there is no electron in the well. It is nevertheless a valid solution.
2. $\sin ka = 0$. $\sin ka$ can be zero if $ka = n\pi (n = 0, 1, 2 \ldots)$. The condition $ka = n\pi$ has the effect of quantising k.

$$k = \frac{2\pi}{\lambda} \quad ka = n\pi = 2\pi a/\lambda$$

so that $\lambda = 2a/n$.

This means that the particle can only exist as standing waves inside the well, as shown in figure 3.7.

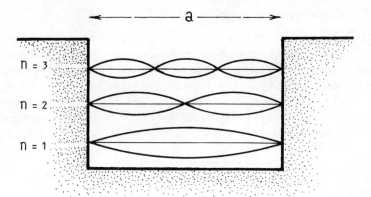

Figure 3.7 First three permitted energy states of a particle in a potential well.

n also quantifies the energies of the standing waves

$$k^2 = 2mE/\hbar^2$$

$$k^2 a^2 = n^2 \pi^2 = \frac{2mE}{\hbar^2} a^2$$

therefore

$$E = \frac{n^2 \pi^2 \hbar^2}{2ma^2} = \frac{n^2 h^2}{8ma^2}$$

This gives the energy corresponding to the values of n which the electron may have in the well. Conversely, n defines the energy of a quantised level.

The solution of the wave equation has shown that the energies in this system must be quantised. The wave equation can also provide the normalised wave functions which correspond to valid solutions.

The general solution $\psi = A \sin kx$ is not a normalised wave function, but this presents no difficulty as it is known that the particle is constrained to exist inside the potential well.

$$\int_0^a \psi \psi^* = 1$$

$$\int_0^a A^2 \sin^2 kx = 1$$

$$A^2 \left[\frac{x}{2} - \frac{\sin 2kx}{4k} \right]_0^a = A^2 \left[\frac{a}{2} \right] = 1$$

$$A = \sqrt{\frac{2}{a}}$$

therefore

$$\psi = \sqrt{\frac{2}{a}} \sin kx = \sqrt{\frac{2}{a}} \sin\left(\frac{n\pi x}{a}\right)$$

In this solution we have been able to find the wave function corresponding to a particle trapped inside a potential well and have obtained an expression for its permitted energy levels. This model will be found to be a good approximation to the behaviour of an electron inside a metal. It could also be used to describe an electron trapped at a potential well (defect or impurity site) in a semiconductor or insulator.

These are only a few examples of the techniques of wave mechanics. One could, for instance, consider a particle which is constrained to exist only on a ring. Again, the permitted energy states would be found to be quantised. An electron orbiting the nucleus of a hydrogen atom can be treated as a particle on a three-dimensional ring where there is an attractive potential

$$V(r) = \frac{Ze^2}{4\pi\epsilon_r \epsilon_0 r}$$

To handle this problem effectively, the wave equation is transformed into spherical polar coordinates. A solution of the equation gives rise to the three quantum numbers n, l and m. Figure 3.8 shows some details of the 1s, 2s and $2p_z$ orbitals of the hydrogen atom.

Two coupled potential wells

The next step is to investigate a system where two potential wells are connected via a barrier which permits a measure of quantum mechanical tunnelling (figure 3.9). One can undertake an exhaustive solution and, provided there are no mistakes in algebra, a solution will be obtained. There is an alternative technique which is frequently used by theoretical chemists during the solution of the wave equation for molecules.

It is assumed that the wave function of the system is a linear combination of the wave functions in each of the wells when isolated from each other.

$$\psi = C_1 \psi_1 + C_2 \psi_2$$

In this approach the values of C_1 and C_2 are altered until the energy of the system is a minimum. The energy of the system is then given by (Cumper, 1966)

$$E = \frac{\int \psi [H] \psi^*}{\int \psi \psi^*}$$

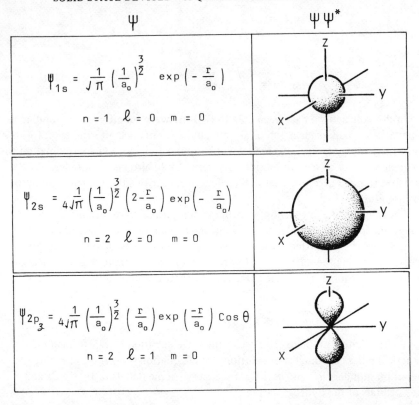

Figure 3.8 ψ and $\psi\psi^*$ for the 1s, 2s and $2p_z$ orbitals of hydrogen.

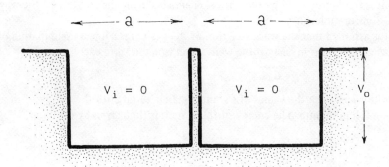

Figure 3.9 Two potential wells connected via a barrier.

therefore

$$E = \frac{\int (C_1\psi_1 + C_2\psi_2)[H](C_1\psi_1 + C_2\psi_2)}{\int (C_1\psi_1 + C_2\psi_2)^2}$$

$$= \frac{C_1^2 \int \psi_1[H]\psi_1 + 2C_1C_2 \int \psi_1[H]\psi_2 + C_2^2 \int \psi_2[H]\psi_2}{C_1^2 \int \psi_1^2 + 2C_1C_2 \int \psi_1\psi_2 + C_2^2 \int \psi_2^2}$$

For shorthand, $\int \psi_i [H] \psi_j$ is written as H_{ij} and is called the *exchange integral*

$\int \psi_i \psi_j$ is written as S_{ij} and is called the *overlap integral*

Therefore

$$E = \frac{C_1^2 H_{11} + 2C_1C_2 H_{12} + C_2^2 H_{22}}{C_1^2 S_{11} + 2C_1C_2 S_{12} + C_2^2 S_{22}}$$

As written, the energy E is going to depend on the exact values of the constants C_1 and C_2. These can be treated as variables so that $E = E(C_1, C_2)$ can be partially differentiated with respect to each. Since the system at equilibrium will be at the lowest possible energy, the values of C_1 and C_2 at E_{min} can be calculated.

Differentiating E with respect to C_1 gives

$$\frac{\partial E}{\partial C_1} (C_1^2 S_{11} + 2C_1C_2 S_{12} + C_2^2 S_{22}) + 2(C_1 S_{11} + C_2 S_{12})E = 2(C_1 H_{11} + C_2 H_{12})$$

Since $\partial E/\partial C_1 = 0$ at E_{min}, this reduces to

$$C_1(H_{11} - ES_{11}) + C_2(H_{12} - ES_{12}) = 0$$

Similarly, differentiating E with respect to C_2 gives

$$C_1(H_{21} - ES_{21}) + C_2(H_{22} - ES_{22}) = 0$$

These are called *secular equations*. In order for there to be non-trivial solutions ($C_1 = C_2 = 0$), the determinant made up of the coefficients of the secular equation must be zero

$$\begin{bmatrix} H_{11} - ES_{11} & H_{12} - ES_{12} \\ H_{21} - ES_{21} & H_{22} - ES_{22} \end{bmatrix} = 0$$

This has been written in a very general form. One would generally start with normalised atomic wave functions so that $S_{11} = S_{22} = 1$. The determinant is then

$$\begin{bmatrix} H_{11} - E & H_{12} - ES_{12} \\ H_{21} - ES_{21} & H_{22} - E \end{bmatrix} = 0$$

Provided that the two atomic wave functions are identical (and they need not be — for example, HCl), then $H_{21} = H_{12}$, $H_{11} = H_{22}$, and $S_{21} = S_{12}$ so that

$$(H_{11} - E)(H_{11} - E) - (H_{12} - ES_{12})^2 = 0$$

or

$$(H_{11} - E) = \pm (H_{12} - ES_{12})$$

Therefore

$$E_+ = \frac{H_{11} + H_{12}}{1 + S_{12}} \quad \text{corresponding to } C_1 = C_2$$

and

$$E_- = \frac{H_{11} - H_{12}}{1 + S_{12}} \quad \text{which corresponds to } C_1 = -C_2$$

It can be seen that the lowest energy state with two wells has two possible values (this result is analogous to what happens when two electrically resonant circuits are coupled).

It was mentioned previously that this technique was used extensively by chemists for the solution of the wave equation. There are two basic approaches to treating a molecule. Both work on an assumption similar to that used above, namely that any molecule formed from component atoms retains some of the characteristics of the atoms themselves; that is, the wave function of the molecule is a linear combination of atomic wave functions.

The valence bond method essentially places all the components into the molecule, constructs a potential which contains all possible attractions and repulsions, and solves the resulting wave equation. Even for a molecule as simple as hydrogen the algebra is tedious. It is not a very popular method.

The molecular orbital approach uses a variational method identical to that given above. The solutions correspond to different energy levels; electrons are then inserted into these levels. This, of course, assumes that there is negligible interaction between electrons in their final locations, which might at first sight seem a bit dubious. Nevertheless, molecular orbital techniques are very elegant and give excellent results.

The treatment of two coupled potential wells had two possible values of lowest energy. A hydrogen molecule also has two ground states. If one considered a problem with three coupled wells, it would be found that there were three ground state energies. This is also true for a triatomic molecule and can be

WAVE MECHANICS

extended further. In general, if a molecule consists of n atoms, then it will have n ground state levels. Such a system ultimately becomes identical to the assembly of particles which constitute a solid, and approximations for solving Schrödinger's equation for the motion of electrons in solids will be considered in the following chapter.

Problems

3.1. Find the normalisation constants for the following wave functions:

(1) $A \cosh x$ which exists only between $x = 2$ and $x = 5$.
(2) $Ax \exp(-k^2 x^2)$ which exists between $\pm\infty$.
(Hint: see problem 2.8 in the previous chapter.)

3.2. A definition of the momentum operator $[P]$ was given in the text. Use this definition to derive the momentum, P, for the following wave function

$$\psi(x, t) = \tfrac{1}{2}[\exp\{j(kx - \omega t)\} + \exp\{-j(kx + \omega t)\}]$$

3.3. For the case $V = 0$, show that $A \cos kx$ is a solution of the Schrödinger wave equation.

3.4. The transmission coefficient for quantum mechanical tunnelling can be approximated by $4\exp(-2k^*a)$, where k^* is the wave number ($= jk$) in the barrier and a is the barrier width.

In a tunnelling experiment, 10^{12} electrons cm^{-2} s^{-1} approach a potential barrier whose energy is 10 keV. The detector on the exit side of the barrier has a maximum sensitivity of 10^2 electrons cm^{-2} s^{-1}. What is the maximum width of potential barrier through which tunnelling can be detected? The effective mass of electrons in the barrier region is $0.4m_e$. It may be assumed that the energy of electrons is very much less than 10 keV.

3.5. The skin depth, δ, for electromagnetic radiation is defined as the distance where the signal amplitude has been reduced by a factor $\exp(-1)$. For a plane wave, δ is given by

$$(\pi \mu f \sigma)^{-1/2}$$

where f is the frequency, μ is the conductor permeability and σ is the conductivity. The skin depth in copper at 1 MHz is 0.066 mm.

If a transmission probability can be approximated by $4\exp(-2a/\delta)$, then calculate the attenuation in dB (defined as $10\log_{10}$ (transmitted amplitude/incident amplitude)) due to a 0.1 mm thick sheet of copper at 100 MHz.

3.6. A particle encountering a finite thickness potential barrier was treated in the text. Now consider the case of two identical finite thickness barriers, each of width a, separated by a distance b.

(1) Derive an expression for the wave function within the interbarrier region.
(2) Derive an expression for the transmission coefficient, defined as the ratio between the square of the amplitude of the particle emerging from the second barrier and the square of the amplitude of the particle incident on the first barrier.

3.7. The example of a finite thickness barrier which was treated in the text considered a barrier that was repulsive. The particle could either be reflected, it could surmount the barrier if it had sufficient energy or it could tunnel through the barrier. One could conversely consider an attractive barrier, such as an ionised impurity site in a solid (similar to the previous example) and analyse the behaviour of an incident particle which was not necessarily trapped in the well. This could be treated by considering a square attractive well of depth $-V$ situated at $x = 0$ and extending from $x = a$ to $x = -a$. For simplicity, it may be assumed that the potential elsewhere is zero.

Solve Schrödinger's wave equation for the regions inside and outside the barrier. It may be assumed that matter and momentum are conserved at the transitions $x = a$ and $x = -a$.

Calculate a transmission probability as the ratio of the square of the amplitudes of the incident and transmitted waves, and observe the effect of the attractive well.

3.8. The problem of two coupled potential wells was treated on the assumption that the wave function of the system was a linear combination of the independent wave functions of the two wells. Repeat the treatment for the case of a potential well of width a coupled to a well of width $2a$. The depth of both wells are identical.

3.9. On the assumption that the principle of linear combination of independent wave functions can be applied to the problem of a particle within three identical coupled potential wells, derive expressions for the allowed lowest energy states.

3.10. An electron is trapped in the vicinity of an ionised phosphorus impurity atom in silicon. The energy required to move the electron from the ground state to the third excited state can be calculated by

(a) treating it as a hydrogen-like Bohr-orbit problem;
(b) treating it as a particle trapped in a square potential box.

Compare the values calculated by these two methods using the information listed below and comment on their validity.

The Bohr energy for hydrogen is

$$E_{n_1} - E_{n_2} = \frac{me^4}{8\epsilon_0^2 \, \epsilon_r^2 \, h^2} \left[\frac{1}{n_2^2} - \frac{1}{n_1^2} \right]$$

The ionisation potential for hydrogen is 13.6 eV.
The radius of the potential well around the phosphorus ion is 50 Å.
The electron effective mass in silicon is $m_e^* = 0.9\, m_e$.
The relative permittivity of silicon is 12.

Reference

C. W. N. Cumper (1966). *Wave Mechanics for Chemists*, Heinemann, London.

Further Reading

Morten Scharff (1969). *Elementary Quantum Mechanics*, Wiley Interscience, New York.

4

Quantum Theories of Solids

From the foregoing, it can be seen that for n atoms forming a molecule, there will be n energy levels, each slightly different; and as n gets larger, the separation in energy between the levels becomes progressively smaller. It is obvious that as n is increased, the computations involved in a solution of the wave equation become less manageable. Simplifying assumptions must be made. There are essentially two approaches to the problem.

The free electron model considers a solid to be an infinite series of closely spaced energy levels inside a potential well. Electrons are added to the levels and are free to move around inside the well. The interactions between electrons and the solid, and between electrons themselves are ignored. In these respects it is analogous to the molecular orbital approach to chemical bonding. Since the electrons are free to move inside the metal and suffer no interactions, the situation is also similar to the kinetics of an ideal gas and it is not unusual to refer to the assembly of electrons as an *electron gas*.

Although the free electron model has its limitations, it is a very useful theory and is particularly valuable in predicting certain phenomena associated with metals.

The band theory of solids uses many of the concepts of the free electron model. In addition, it considers the crystal lattice of the solid to represent a regular array of periodic potential wells, with which the electrons interact. A formal treatment using this model is usually concerned with the effects of these interactions, which give rise to energy bands and forbidden gaps.

QUANTUM THEORIES OF SOLIDS

The Free Electron Theory of Solids

Figure 4.1 shows a sea of electrons trapped inside a potential well of height E_s. Each electron is free to move inside the well and has no interactions with any other electrons. In order to escape from the well an electron would need to acquire an additional amount of energy, an *activation energy*.

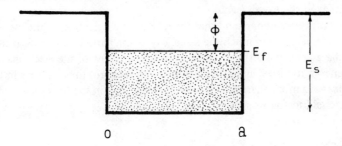

Figure 4.1 A free electron metal.

A molecular orbital approach can be used to construct the energy levels into which electrons can subsequently be placed. Since each electron in the well is free of any interactions, the solution of Schrödinger's wave equation for that electron gives

$$E_n = \frac{n^2 h^2}{8ma^2} = \frac{k^2 h^2}{8\pi^2 m}$$

In chapter 2, several amplitude–frequency plots were presented. In chapter 3, it was mentioned that the square of a wave amplitude is a measure of its energy. The formula for the energy of an electron in a potential well shows that an energy *versus* wave number plot (figure 4.2) describes a parabola in k-space. The levels are discrete but very closely packed.

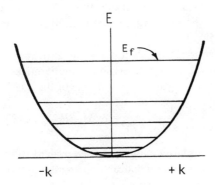

Figure 4.2 A plot of E versus k, showing the energy levels up to E_f.

One can now start to fill the levels, remembering that the Pauli Exclusion Principle must hold at all times. That states that no two electrons with the same quantum numbers can occupy the same energy level. Accordingly, each level contains two electrons, one with spin $+\frac{1}{2}$ and the other with spin $-\frac{1}{2}$. The electrons will first occupy the lowest levels, and then levels of increasing energy until the available supply of electrons is exhausted. The topmost level which is occupied is called the *Fermi level*.

One can intuitively understand that the energy of an electron in the metal depends on temperature in much the same way as the particles in an ideal gas. However, there are some differences. At zero Kelvin, none of the electrons have sufficient energy to move; and at temperatures above zero, the electrons in the upper levels move more energetically than the electrons in the lower levels. The electrons are said to follow a Fermi-Dirac distribution.

Fermi-Dirac statistics

In Fermi-Dirac statistics the probability of an electron having energy E is given by

$$P(E) = \frac{1}{\{\exp[(E - E_f)/kT] + 1\}}$$

It is used when electrons occupy energy levels which are so close together as to be virtually indistinguishable.

Properties of a Fermi-Dirac distribution

The properties of electrons which follow a Fermi-Dirac distribution depend entirely on whether the temperature is at or above zero Kelvin and whether one considers energies above or below the Fermi energy level.

At zero Kelvin (see figure 4.3a), the formula for the distribution confirms that the probability of finding an electron below the Fermi energy is unity and above the Fermi energy is zero. That means that there are no electrons with energy greater than the Fermi energy, which in itself forms the basis of the definition of that level.

For temperatures above zero Kelvin, the exponential part of the formula is finite. Some electrons will occupy levels above the Fermi level (see figure 4.3b) and the probability of finding an electron in one of these levels is

$$P(E) \approx \frac{1}{\exp[(E - E_f)/kT]} = \exp[-(E - E_f)/kT]$$

It can be seen that this latter expression is equivalent to the Maxwell-Boltzmann distribution, which is applicable to an assembly of particles, such as an ideal gas, where the energy levels are not degenerate.

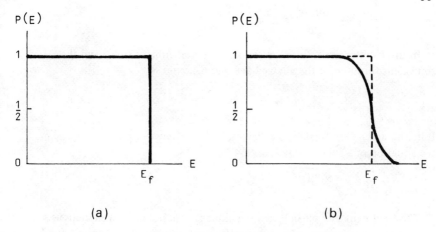

Figure 4.3 (a) A Fermi-Dirac distribution at zero Kelvin. (b) Fermi-Dirac distribution at temperatures above zero Kelvin.

The Maxwell-Boltzmann probability function is given by

$$P(E) = \frac{A}{(2\pi mkT)^{3/2}} \exp(-E/kT)$$

Thus a Fermi-Dirac distribution has a Maxwell-Boltzmann tail at temperatures above zero Kelvin. It is worth noting that only the electrons within this tail contribute to the specific heat of the solid. Fermi-Dirac electrons do not. This accounts for the observations that the Dulong and Petit law (specific heat at constant volume for metals = 25 kJ kmol^{-1} K^{-1}) does not hold at low temperatures.

It can be seen that the free electron model provides energy levels into which electrons may be 'poured'. At zero Kelvin, these occupy all levels up to the Fermi level. Above zero Kelvin, the uppermost electrons are excited.

With this in mind, one might then ask whether it is possible to calculate the number of electrons occupying levels within a specified energy range. The response to this question depends on two independent factors. The first is the total number of energy levels per unit volume, $S(E)$, which are available for possible occupation by electrons. This is usually called the *density of states*. The second is the probability, $P(E)$, of an electron having sufficient energy to occupy an available level. Thus we can say that the number of electrons between E and $E + dE$ is

$$n(E)\, dE = S(E)\, P(E)\, dE$$

$P(E)$ can be obtained from the Fermi-Dirac formula.

The derivation of $S(E)$

For a particle trapped in a one-dimensional potential well of length a

$$k^2 a^2 = \frac{2mEa^2}{\hbar^2} = n^2 \pi^2$$

In three dimensions, this represents a potential cube of side a and the equivalent formula for the allowed energies becomes

$$\frac{8mEa^2}{h^2} = n_x^2 + n_y^2 + n_z^2$$

If n_x, n_y and n_z are treated as variables, then the above expression represents the equation of a sphere of radius

$$R = \sqrt{\frac{8mEa^2}{h^2}}$$

The treatment of a particle in a one-dimensional potential well demonstrated that n was a quantum number which defined discrete allowable energy states. Thus any set of values of n_x, n_y, n_z defines a unique energy cube. The problem of finding the number of energy states between E and $E + dE$ reduces to a problem of finding the number of energy cubes between R and $R + dR$ on a spherical surface. However, as only positive values of n_x, n_y and n_z are allowed, one need only consider the positive octant of a sphere (figure 4.4).

The number of cubes between R and $R + dR$ in the positive octant is equivalent to the volume between R and $R + dR$

$$\frac{1}{8}[4\pi R^2 \, dR] = \frac{4\pi}{8} \frac{8mE\,a^2}{h^2} \frac{1}{2}\left[\frac{8mE\,a^2}{h^2}\right]^{1/2} E^{-1/2} \, dE$$

Thus the number of electron states between E and $E + dE$ is

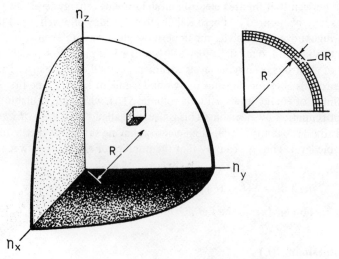

Figure 4.4 Energy cubes on an energy surface.

QUANTUM THEORIES OF SOLIDS

$$\frac{4\pi}{8}\left[\frac{8m}{h^2}\right]^{3/2} E^{1/2}\, V\, dE$$

where $V = a^3$. Note that there are two electrons per energy cube, since the spin quantum number has two states. The number of electron states per unit volume between E and $E + dE$

$$S(E)\, dE = \frac{\pi}{2}\left[\frac{8m}{h^2}\right]^{3/2} E^{1/2}\, dE$$

The number of occupied states per unit volume between E and $E + dE$

$$n(E)\, dE = \frac{\pi}{2}\left[\frac{8m}{h^2}\right]^{3/2} E^{1/2} \frac{1}{\{\exp\left[(E - E_f)/kT\right] + 1\}}$$

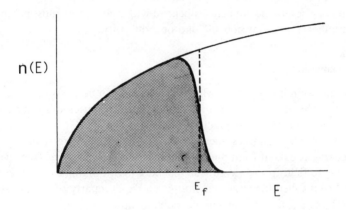

Figure 4.5 The density of occupied states as a function of energy.

The expression for density of states can be immediately utilised in several ways. Since the Fermi–Dirac distribution at zero Kelvin gives a probability of unity of finding an electron at any energy less than the Fermi level, one should be able to calculate the total number of electrons with energies up to E_f.

$$N = \int_0^{E_f} n(E)\, dE = \int_0^{E_f} \frac{\pi}{2}\left[\frac{8m}{h^2}\right]^{3/2} E^{1/2}\, dE$$

$$= \frac{\pi}{2}\left[\frac{8m}{h^2}\right]^{3/2} \frac{2}{3} E_f^{3/2}$$

Alternatively, since one already knows N, the above expression could be used to calculate E_f.

$$E_f = \left(\frac{3N}{\pi}\right)^{2/3} \frac{h^2}{8m} = \left(\frac{3N}{8\pi}\right)^{2/3} \frac{h^2}{2m}$$

Classical mechanics would have $E_f = 0$ at zero Kelvin.

The average energy of electrons at zero Kelvin (given by the sum total energy of all electrons divided by the total number of electrons) can also be calculated as

$$\frac{1}{N}\int_0^{E_f} n(E)E\, dE = \frac{1}{N}\int_0^{E_f} S(E)\,P(E)E\, dE = \frac{1}{N}\int_0^{E_f} S(E)E\, dE$$

$$= \left[\frac{\pi}{2}\left(\frac{8m}{h^2}\right)^{3/2}\frac{E_f^{3/2}}{3/2}\right]^{-1}\int_0^{E_f}\frac{\pi}{2}\left(\frac{8m}{h^2}\right)^{3/2}E^{3/2}\, dE = \frac{3}{5}E_f$$

Again, this is a non-classical result.

The free electron model of solids provides an excellent theoretical framework for treating several phenomena associated with metals. The most notable examples are thermionic emission and photoconductivity.

Thermionic emission

Thermionic emission is the process whereby electrons can be removed from a metal by the application of heat. It forms the basis of operation of most electronic valves and is an essential feature in all cathode ray tubes.

In order for an electron to escape from a metal it must acquire at least sufficient energy to take it from the Fermi level to the vacuum level (the level corresponding to field-free space outside the metal). This minimum requirement (see figure 4.6) is called the *work function*, ϕ, and is a property of any metal. For the alkali metals, like sodium and potassium, and the alkaline earth metals, like barium, it is of the order of 2.25 eV. For most other metals, it is at least twice that value.

Since all electrons which are likely to acquire sufficient energy to overcome the potential ϕ and emerge from the metal are Maxwell-Boltzmann electrons, one can use a kinetic theory of gases approach to calculate thermionic current as a function of temperature for a given work function.

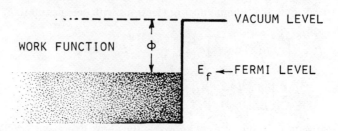

Figure 4.6 A free electron metal.

In the derivation, which is moderately complicated, one calculates the number of electrons per unit area within an energy range which approach the inner surface of the metal in one second. Only those electrons whose energy is greater than the work function are capable of escaping. However, from our previous treatment of the interactions of particles with potential steps, we know that electrons may be reflected at the potential discontinuity. Thus, any expression for the number of electrons which emerge from the metal must contain a term $(1 - \rho)$, where ρ is a reflection coefficient. The thermionic current is then

$$J = \frac{4\pi mek^2}{h^3} T^2 (1 - \rho) \exp\left(-\frac{\phi}{kT}\right) = A(1 - \rho) T^2 \exp\left(-\frac{\phi}{kT}\right)$$

A is called the *Richardson* or *Richardson–Dushman constant*. It has a value 120 A K m^{-2}. It can be seen that the current has an exponential dependence on the work function.

In practice, tungsten is used whenever a thermionic emission source is required. It is one of the few metals capable of withstanding the very high temperatures which are necessary to get a reasonable emission current. Sources based on tungsten alone are called *bright emitters* because of the operating temperature. They are notably short lived, even when the tungsten contains additives such as thorium to improve its mechanical properties. The reliability of electronic valves and cathode ray tubes has been much enhanced by the introduction of *dull emitters*. These generally consist of a tungsten backing which is coated with various alkaline earth oxides. It has been found that the current depends only on the work function of the surface material. As the alkaline earth oxides have very low work functions, it is possible to obtain high emission currents at temperatures as low as 600°C. Figure 4.7 shows the basic components of a triode valve which uses an indirectly heated, dull-emitting cathode.

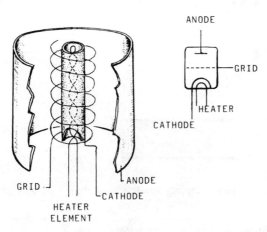

Figure 4.7 Cutaway view of a triode valve with circuit symbol shown as inset. Note the indirectly heated cathode.

Field-enhanced emission

Truly abrupt discontinuities rarely occur in nature. The potential barrier at the surface of the metal is probably closer to that shown in figure 4.8.

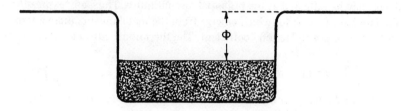

Figure 4.8 A free electron potential well with non-abrupt barriers.

Let us suppose that an electron moves out from the bulk of the electron gas towards the potential barrier. As it approaches the metal surface, the atoms in its vicinity become polarised as a result of charge imbalance at the change of medium. The electron consequentially 'sees' a reflection of itself called an *image charge*. The image charge outside the metal exerts a Coulombic attraction on the electron inside (see figure 4.9).

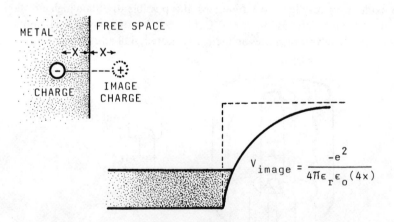

$$V_{image} = \frac{-e^2}{4\pi\epsilon_r\epsilon_o(4x)}$$

Figure 4.9 Band bending due to image charge attraction.

Thus the image potential reduces the reflection coefficient of the escaping electrons and also reduces the apparent work function. The effect on work function can be easily quantified.

If an electric field $\mathcal{E}$ is applied, then the potential at any point x is:

$$V(x) = \frac{-e^2}{4\pi\epsilon_r \epsilon_0 (4x)} - e\mathcal{E}x$$

The effect of this field on the barrier height can be seen in figure 4.10.

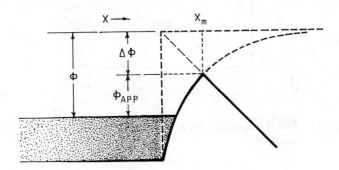

Figure 4.10 Barrier lowering due to image charge effects.

The potential is a maximum when

$$dV(x)/dx = 0$$

or

$$x_m = \frac{1}{2}\left[\frac{e}{4\pi\epsilon_r \epsilon_0 \mathcal{E}}\right]^{1/2}$$

Substituting this into the expression for $V(x)$ gives the value of potential at x_m

$$V(x_m) = \frac{-e}{4\pi\epsilon_r \epsilon_0} (e\mathcal{E})^{1/2} = -\Delta\phi$$

$\Delta\phi$ is the amount by which the apparent work function is less than the absolute work function. The emission current is given by

$$J = A(1-\rho) T^2 \exp\left[-\frac{(\phi - \Delta\phi)}{kT}\right] = J_0 \exp[\Delta\phi/kT]$$
$$= J_0 \exp[B\mathcal{E}^{1/2}/kT]$$

$$B = \frac{e^{3/2}}{4\pi\epsilon_r \epsilon_0}$$

The appearance of the electric field strength in the exponential term confirms that barrier lowering is responsible for the Schottky effect. The emission current from the cathode of a thermionic diode is found to be proportional to the square root of the anode potential (see figure 4.11).

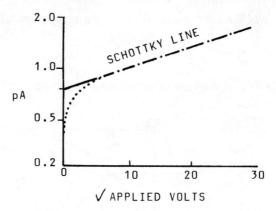

Figure 4.11 A plot of current *versus* the square root of anode voltage for tungsten.

Field emission (Fowler-Nordheim tunnelling)

It can be seen from figure 4.10 that the application of an electric field reduces the effective barrier height. As such a field is increased the width of the potential barrier, which separates the electron gas from the outside world, is significantly reduced (see figure 4.12).

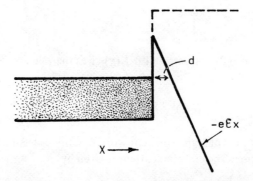

Figure 4.12 Fowler-Nordheim tunnelling: band bending at very high electric fields.

As the field is increased further, there will come a point where electrons are able to pass through the barrier by quantum mechanical tunnelling. This phenomenon is called *cold field emission* or *Fowler-Nordheim tunnelling* (Fowler and Nordheim, 1925). The process requires very high fields. For example, a 10 Å barrier width would require a field of the order of 10^9 V m^{-1}. It is known from simple theory that the electric field is inversely proportional to the square of distance. Thus, it should be possible to obtain extremely high fields in the vicinity of the tip of a metal whisker where the radius is very small.

This forms the basis of the field emission microscope (see figure 4.13), which was one of the first instruments to examine metal surfaces in atomic detail. Electrons emitted from atomic irregularities at the end of the metal whisker will travel radially along the field lines and strike the film, giving a reproduction of these irregularities. The rate of electron emission will be different at different crystal faces and will be affected by the presence of adsorbed materials on the surface. Thus, the technique has been used extensively as a research tool in the examination of the catalytic action of metals such as tungsten and platinum, and the effects of poisons which render catalysts ineffective. Sulphur is an example of an element which is present in petroleum and which can poison noble metal catalysts used in cracking and refining.

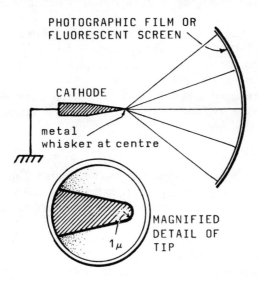

Figure 4.13 Schematic view of a field emission microscope.

The field ion microscope (Muller and Tsong, 1969) is almost identical in design to the emission microscope, but operates at both higher gas pressure and voltage. In this arrangement the screen forms the cathode. The metal whisker becomes the anode and is held at ground potential. Ionised gas particles travel along the field lines towards the anode. The radius of the microscope is several mean free paths at the operating gas pressure. Thus, energetic electrons will collide several times with the intervening gas, causing extensive ionisation. Positive gas ions are then attracted back towards the cathode and if they gain enough energy, they too can cause further ionisation. Thus, a single emerging electron can give rise to an avalanche of positive ions which strike the photographic film at the cathode. As positive ions are much heavier than electrons, the resulting photo-image is more intense than would be possible with a field emission microscope.

Photo-electric effect

The photo-electric effect is a phenomenon whereby the energy of individual photons of light is transferred to electrons inside a metal, giving them sufficient energy to overcome the potential barrier. At first sight it would appear to be impossible to account for the photo-electric effect using the free electron model. In order for a truly free electron in the metal to absorb a photon, its position must be fixed in space during the process of absorption. This would contravene the Heisenberg Uncertainty Principle. One can take the view that only electrons held close to the surface by image charge and other attractions are sufficiently located to allow an energy transfer to take place. There is, nevertheless, a contribution from the bulk of the metal and this is largely due to the fact that the free electron model is not strictly valid. If, however, these theoretical shortcomings are ignored, the free electron model provides an excellent framework for treating the photo-electric effect.

The magnitude of a photo-electric current depends on the intensity of the incident light. The effect of optical excitation frequency is summarised by the Einstein equation (Einstein, 1905)

$$hf = e\phi + \tfrac{1}{2}mv^2$$

The total energy of the photon is transferred to the electron. This first provides the electron with sufficient energy to rise from the Fermi level to the vacuum level. Any surplus then appears as kinetic energy in the electron which emerges from the metal.

Since the frequency of incident radiation is known and the kinetic energy of electrons can be measured, the application of the Einstein equation to photo-electric measurements is an ideal method for determining the work function of metals. In this context, it should be noted that photo-electric measurements provide a very accurate value of ϕ. Work function can also be determined using thermionic emission methods. However, there is a statistical spread in the values owing to the fact that the high temperatures result in a Maxwell-Boltzmann spread of energies about the Fermi level.

Band Theory of Solids

The free electron model is very successful, as far as it goes. It has, however, one major deficiency: in the simplistic approach, the periodic nature of the crystal lattice of most solids is ignored. Accordingly, the model is unable to account for phenomena which are a direct result of electron-lattice interactions. The band theory of solids attempts to overcome this problem. As an electron moves through a crystal lattice it comes under the influence of the potential well of each atom or ion in its track. The wave function of an electron in the solid should therefore reflect this state of affairs. As a simple approximation, one could express this wave function as a linear combination of the electron wave functions associated with each isolated atom (ion) of the solid.

Bloch (1928) approached the problem from a slightly different standpoint. He suggested that the electrons of a solid belonged to the crystal rather than to individual atoms. On this basis, a theorem was proposed. If a potential is periodic such that

$$V(x) = V(x + a)$$

then solutions of Schrödinger's wave equation must yield wave functions of the form

$$\psi = \exp(\pm jkx)\, U_k(x)$$

$U_k(x)$ is a function which takes account of the periodicity of the lattice. It is called a *Bloch function* and has the property

$$U_k(x) = U_k(x + a)$$

In tackling this problem, simplifications must be made. The mathematics would be almost impossible if one were to use realistic periodic potentials. The Kronig-Penney model represents the most popular approach (Kronig and Penney, 1931). It analyses the motion of electrons in a one-dimensional array of square potentials of width a, separated from each other by potential barriers of height V_0 and width b (see figure 4.14).

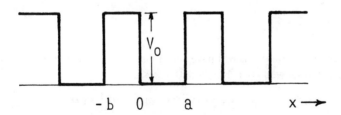

Figure 4.14 An ideal periodic potential.

It is quite clear from figure 4.14 that there are two distinct regions where Schrödinger's wave equation must simultaneously have solutions.

In the potential well (that is, where $0 < x < a$), because $V = 0$

$$\frac{d^2\psi}{dx^2} + \frac{2me}{\hbar^2}\psi = 0 \tag{4.1}$$

In the potential barrier (that is, where $-b < x < 0$), because $V = V_0$

$$\frac{d^2\psi}{dx^2} + \frac{2m}{\hbar^2}(E - V_0)\psi = 0 \qquad (4.2)$$

It is assumed throughout the derivation that $V_0 > E$.

In order to facilitate the mathematical manipulation, it is useful to define two real quantities α and β

$$\alpha^2 = \frac{2mE}{\hbar^2}$$

$$\beta^2 = \frac{2m(V_0 - E)}{\hbar^2}$$

Note that α is directly related to the energy, E, of the electron.

It will be remembered from chapter 3 that we know the form of a solution of Schrödinger's wave equation. This is then combined with boundary conditions in order to obtain an exact solution. According to Bloch's theorem, the wave equation within a periodic potential has solutions of the form $\exp(jkx)U_k(x)$. If this expression is entered into the wave equation for the two distinct regions, equations (4.1) and (4.2) above, then the exponential term, $\exp(jkx)$, will cancel throughout. The result will be a new pair of equations in $U_k(x)$, written as U for short.

In the potential well (that is, where $0 < x < a$)

$$\frac{d^2 U}{dx^2} + 2jk\frac{dU}{dx} + (\alpha^2 - k^2)U = 0 \qquad (4.3)$$

In the potential barrier (that is, where $-b < x < 0$)

$$\frac{d^2 U}{dx^2} + 2jk\frac{dU}{dx} - (\beta^2 + k^2)U = 0 \qquad (4.4)$$

These differential equations will have solutions of the following form.

In the potential well (that is, where $0 < x < a$)

$$U_{k_1} = A\exp[j(\alpha - k_1)x] + B\exp[-j(\alpha + k_1)x] \qquad (4.5)$$

where k_1 is the wave number in this region.

In the potential barrier (that is, where $-b < x < 0$)

$$U_{k_2} = C\exp[(\beta - jk_2)x] + D\exp[-(\beta + jk_2)x] \qquad (4.6)$$

where k_2 is the wave number in this region.

The following conditions can be applied to solve for A, B, C and D.

(I) Because of continuity of matter and momentum

$$U_{k_1}(0) = U_{k_2}(0) \qquad \frac{dU_{k_1}(0)}{dx} = \frac{dU_{k_2}(0)}{dx}$$

(II) Because of the periodic properties of Bloch functions

$$U_{k_1}(a) = U_{k_2}(a) \qquad \frac{dU_{k_1}(a)}{dx} = \frac{dU_{k_2}(a)}{dx}$$

With these it should be possible, although quite tedious, to obtain values for A, B, C and D. However, for our purposes this is not at all necessary. It is much more important to know the energy levels of the electrons in this periodic lattice which correspond to satisfactory solutions of the wave equation.

The simultaneous equations, (4.5) and (4.6), will have solutions only if the coefficient determinant is zero.

$$\begin{bmatrix} A & B \\ C & D \end{bmatrix} = 0$$

This means that there will only be solutions if

$$\frac{\beta^2 - \alpha^2}{2\alpha\beta} \sinh \beta b \sin \alpha a + \cosh \beta b \cos \alpha a = \cos k(a+b) \qquad (4.7)$$

In order to simplify this, Kronig and Penney (1931) suggested that the barriers could be treated as delta functions. In this case the barrier height V_0 is allowed to tend towards infinity as the barrier width tends towards zero, so that the product bV_0 remains finite. In this case, equation (4.7) simplifies to

$$\frac{mV_0 b \sin \alpha a}{\hbar^2 \alpha} + \cos \alpha a = \cos ka \qquad (4.8)$$

This can be rewritten as

$$\frac{P \sin \alpha a}{\alpha a} + \cos \alpha a = \cos ka \qquad (4.9)$$

$$P = \frac{mV_0 ba}{\hbar^2}$$

Note that equation (4.9) is not a solution of the wave equation of an electron in a periodic lattice. It is a condition for an electron to have a solution. Since the conditions contains α, which itself is a function of energy E, one should be able to find the permitted values of E which correspond to satisfactory solutions.

Figure 4.15 shows a graph of the left-hand side of equation (4.9) *versus* αa for $P = 100$. One can see how the amplitude decreases and period increases with increasing αa. It must be remembered that equation (4.9) represents an identity. It is only true when the left-hand side is equal to $\cos ka$. However, $\cos ka$ can only have values between +1 and −1. Thus, the identity restricts the permitted values which correspond to a satisfactory solution.

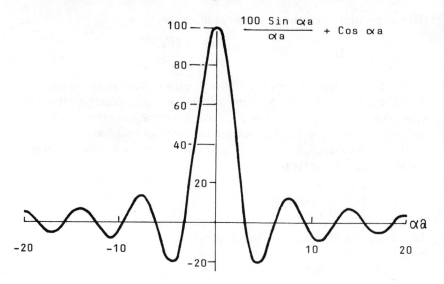

Figure 4.15 A plot of $\dfrac{\sin \alpha a}{\alpha a} + \cos \alpha a$ versus αa for $P = 100$.

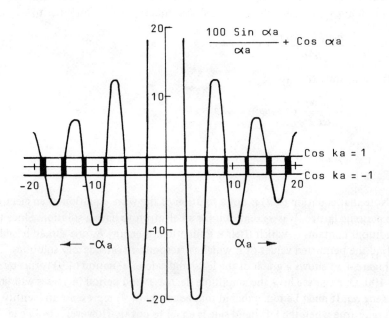

Figure 4.16 A close-up of figure 4.15, showing the regions over which $\dfrac{\sin \alpha a}{\alpha a} + \cos \alpha a = \cos ka$.

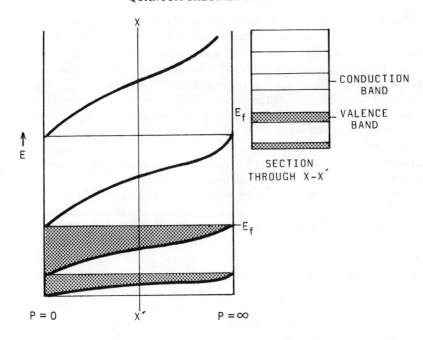

Figure 4.17 A schematic diagram of the relationship between E and P: the inset shows a section x-x' which is identical to the usual representation of valence and conduction bands.

This restriction means that Schrödinger's wave equation will only have valid solutions in the dark areas shown in figure 4.16. In other words, an electron moving in a periodic potential will be able to occupy only bands of energy levels. The permitted bands are separated from each other by forbidden gaps, regions where the left-hand side of equation (4.9) is less than -1 or greater than $+1$.

Equation (4.9) also determines many of the properties of the permitted energies. Because of the nature of the function, the separation between bands and the width of the bands themselves increase with increasing αa. The constant P is a measure of the area of the barriers in the periodic potential. As P gets larger the electrons become more tightly bound to a particular potential well. This is shown schematically in figure 4.17.

If $P = 0$, then we have the condition $\alpha a = ka$. This is identical to the situation in the free electron model. There is no binding to the lattice. As P becomes larger the electron becomes more tightly bound to an individual potential well. In the extreme, it behaves as an isolated particle in a box. For intermediate values of P one obtains bands separated by gaps. The more familiar representation of energy bands can be seen in the cross-section shown as an inset in figure 4.17.

Since $\alpha^2 = 2mE/\hbar^2$, it should be possible to relate the identity given by equation (4.9) to an E–k parabola, similar to that used in the Free Electron model. The left-hand side of figure 4.18 shows an exaggerated view of figure 4.16 which has been rotated through 90°. The vertical axis, αa, is a direct measure of energy. The diagram on the right-hand side is an E–k energy parabola, where the k-axis is divided in units of π/a. The regions of $[\sin \alpha a]/\alpha a + \cos \alpha a$ which lie between +1 and −1 are marked as solid lines on the parabola. These represent the regions of permitted energy where the electron in the periodic lattice behaves as if it were a free electron. The regions of forbidden energy are shown as broken lines on the parabola. The reason why the solid lines are flattened out at their extremes will become clear shortly.

The bands of permitted energy are usually called Brillouin zones (Brillouin, 1953). Since $E = k^2\hbar^2/2m$ in the Free Electron model and E was defined as $\alpha^2\hbar^2/2m$ in the Band Theory, it is obvious that figure 4.18 represents only half the story. There will also be positive energies for negative values of α or k. This is shown in figure 4.19.

This figure is rather cumbersome to deal with, but it can be simplified. Since $\cos(ka) = \cos(ka + 2n\pi/a)$, $n = 1, 2, \ldots$, it should be possible to represent the diagram in reduced k-space. A reduced k-space construction could be considered to be something like an exercise in folding paper. If figure 4.19 were drawn on tracing paper, then there could be a series of folds at π/a, $2\pi/a$ etc. The diagram would first be folded back at y and y', the outer extremes of the

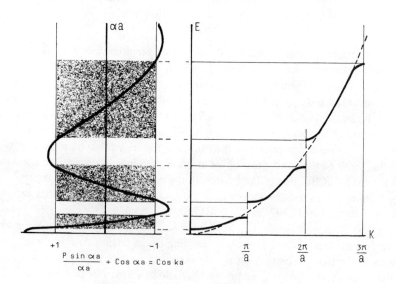

Figure 4.18 The relationship between $\dfrac{\sin \alpha a}{\alpha a} + \cos \alpha a = \cos ka$ and the E–k diagram.

first Brillouin zones. The points $\pm 2\pi/a$ would coincide with $k = 0$, which would represent the mid-points of the second and third Brillouin zones. A further pair of outward folds at $\pm 2\pi/a$ would bring $+3\pi/a$ into coincidence with π/a and $-3\pi/a$ with $-\pi/a$. The construction for positive values of k is shown in figure 4.20a. Figure 4.20b shows the complete reduced k-space diagram.

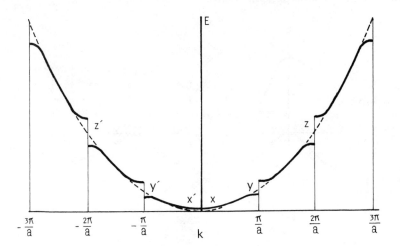

Figure 4.19   An E-k diagram for positive and negative values of k.

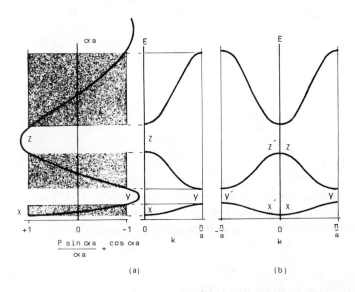

Figure 4.20 (a) The construction of an E-k diagram in reduced k-space. (b) A reduced E-k diagram for positive and negative values of k.

The exact shape of the reduced k-space diagram depends entirely on the crystal lattice. If the value of a, the width of the potential well, is analogous to the spacing between crystal planes, then it will be different for (100), (110), (111) etc. planes. Figure 4.21 shows the reduced k-space energy parabolas in the region of the highest filled and lowest empty levels for (a) silicon and (b) gallium arsenide (Cohen and Bergstresser, 1960).

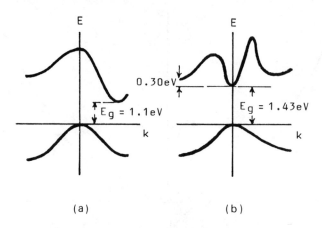

Figure 4.21 Schematic k-space diagrams for (a) silicon, (b) gallium arsenide.

The velocity of an electron in an energy band

It was mentioned earlier that the reason for the flattening of the permitted energies in the E-k parabola would be explained. It will now be shown that this is related to the velocity of the electron within a Brillouin zone.

The velocity of a particle is equal to the group velocity of a wave

$$v = \frac{d\omega}{dk} = \frac{d(2\pi f)}{dk} = \frac{d}{dk}\left[\frac{2\pi h f}{h}\right]$$

$$= \frac{1}{\hbar}\frac{dE}{dk}$$

The electron velocity will only be zero when dE/dk is zero. Thus, the edges of the permitted energies on the E-k diagram must be flat. If this were not the case, the electron would have velocity at the edge of a Brillouin zone and would therefore escape into the forbidden region.

The effective mass of an electron in an energy band

The effective mass of an electron in an energy band can be calculated by estimating the influence which an electric field might exert upon it.

A field is applied to an electron with wave number k during time dt. The energy gain

$$dE = e\mathscr{E}v\,dt = \frac{e\mathscr{E}}{\hbar}\frac{dE}{dk}dt$$

$$= \frac{dE}{dk}dk$$

therefore

$$dk = \frac{e\mathscr{E}}{\hbar}dt$$

$$\frac{dk}{dt} = \frac{e\mathscr{E}}{\hbar}$$

The acceleration experienced by the electron

$$\frac{dv}{dt} = \frac{d}{dt}\left[\frac{1}{\hbar}\frac{dE}{dk}\right]$$

$$= \frac{1}{\hbar}\frac{d}{dk}\left[\frac{dE}{dk}\right]\frac{dk}{dt} = \frac{1}{\hbar}\frac{d^2E}{dk^2}\frac{dk}{dt} = \frac{e\mathscr{E}}{\hbar^2}\frac{d^2E}{dk^2}$$

and also

$$\frac{dv}{dt} = \frac{\text{force}}{m^*} = \frac{e\mathscr{E}}{m^*} = \frac{e\mathscr{E}}{\hbar^2}\frac{d^2E}{dk^2}$$

therefore

$$m^* = \hbar^2/[d^2E/dk^2]$$

At the bottom of a band the change of energy with k is changing in a positive sense. The mass of the electron will have a positive sign. On the other hand, at the edge of a Brillouin zone the rate of change of E with k changes in a negative sense. In this case the electron will have negative mass. This is a way of saying that electrons exist at the bottom of bands and holes exist at the top of bands. At the points of inflection the mass of the electron becomes infinite.

It is worth taking special note of the nature of d^2E/dk^2. It is determined by the shape of the E-k parabola in reduced k-space. Thus, an electron in a very steep parabola will have a smaller effective mass than an electron in one which is less steep. This has particular significance for electrons in the conduction bands of gallium arsenide, as will be seen later.

The Band Theory of solids can now be used to distinguish between metals, insulators and semiconductors. The metallic state can arise as a result of an incomplete filling of the highest occupied energy band. It can also be due to a filled band and an empty band whose energies on an E-k diagram are seen to overlap. In this case there is nothing which prevents electron transfer between the two bands. The forbidden gap energy separation between the lowest empty or conduction band and the highest filled or valence band determines whether a material is an insulator or a semiconductor. This is of the order of about 1 eV for semiconductors (1.1 eV for silicon, 0.7 eV for germanium). For insulators it is much larger (10 eV for sodium chloride).

Above all, it is the crystal lattice dimensions and symmetry which determine the band structure and hence the electrical properties of many solids. Sometimes it only requires small variations in temperature (which bring about only minor changes in crystal dimension) to cause significant changes in electrical properties. Vanadium pentoxide (V_2O_5) undergoes a metal-to-semiconductor transition at low temperatures. There is a seven orders of magnitude change in resistivity. Barium titanate ($BaTiO_3$) undergoes a similar but smaller change at about $100°C$ and is often used as a positive coefficient thermistor. Tin undergoes a metal-to-semiconductor transition at about $-30°C$. Unlike the metal, grey tin is a powdery material with no mechanical strength. It has been suggested that this transition, sometimes called 'tin-pest', was largely responsible for the collapse of Napoleon's Russian Campaign in the winter of 1812.

Problems

4.1. The energy of a particle in a three-dimensional rectangular well is given by

$$\frac{h^2}{8m}\left[\frac{n_x^2}{a_x^2} + \frac{n_y^2}{a_y^2} + \frac{n_z^2}{a_z^2}\right]$$

The separation between (111) planes in a simple cubic crystal is 2.5 Å. A particle (mass = m_e) is trapped in a rectangular box within the crystal. The potential box has the following dimensions: the length is defined by the separation between the $(\bar{1}10)$ and $(1\bar{1}0)$ planes, the width is equivalent to d_{110}, and the height is equivalent to d_{001}. Calculate the value of the ground state energy.

4.2. Aluminium has a density of 2700 kg m^{-3} and an atomic weight of 26.9. Each atom contributes three conduction electrons. Calculate the value of the Fermi energy at zero Kelvin (mass = m_e).

4.3. The lowest energy level for conduction electrons in sodium lies 5.38 eV below the vacuum level. The work function is 2.28 eV. Estimate the total number of electrons per cubic metre below the Fermi level at absolute zero, and calculate their average energy. Assume that $m^* = m_e$.

4.4. The work function of tungsten is 4.55 eV. What temperature would be required in order for an electron to have a 10^{-6} probability of occupying an energy level that would allow it to escape without further activation? Thorium is often added to improve the ductility of tungsten in the manufacture of filaments. It will also improve the emission due to its lower work function (3.4 eV). If a thoriated tungsten filament is heated to 3500 K, what is the probability of an electron having sufficient energy to escape.

Modern thermionic sources use much lower temperatures. The tungsten is coated with barium oxide whose work function is of the order of 1.5 eV. The dramatic reduction in operating temperature can be observed by calculating the condition for a 10^{-6} probability and comparing it with the value calculated for tungsten in the first part of the question.

4.5. At 1500°C a metal filament emits a current of 0.1 A m^{-2}. If the work function is 3.55 eV, calculate the percentage of electrons with $E \geqslant \phi$ which strike the barrier but fail to escape.

4.6. A filament identical to that in problem **4.5** has an anode placed close by. Estimate the magnitude of an applied field which would bring about a twenty-fold increase in current at 1500°C. It can be assumed that $\epsilon_r = 50$.

4.7. A nickel whisker is used in a Fowler–Nordheim tunnelling experiment. The work function of nickel is 5.15 eV and the width of the barrier at E_f is 8 Å. What is the minimum field which is necessary for tunnelling if the experiment is carried out at very low temperatures?

4.8. If an electron at point x in a periodic crystal lattice is described by the wave function $\psi(x) = \exp(jkx)U_k(x)$, where $U_k(x)$ is a Bloch function, then prove that $\psi(x + a) = \psi(x)\exp(jka)$.

4.9. Plot $$\frac{P \sin \alpha a}{\alpha a} + \cos \alpha a$$

for positive values of αa, for $P = 1000$ and $P = 50$.

Estimate the ranges of αa (up to $\alpha a = 20$) over which the condition

$$\frac{P \sin \alpha a}{\alpha a} + \cos \alpha a = \cos ka$$

is satisfied.

4.10. A simple cubic crystal has a unit cell dimension of 2.5 Å. Draw plots of

$$\frac{P \sin \alpha a}{\alpha a} + \cos \alpha a$$

for the motion of an electron in the periodic potential defined by
(1) the (100) planes,
(2) the (110) planes,
(3) the (111) planes.

For each of the above cases, plot E against k for the first permitted set of energies. $P = 10^{12} a$.

References

F. Bloch (1928). *Z. Phys.*, **52**, 555.

L. Brillouin (1953). *Wave Propagation in Periodic Structures*, 2nd edn, Dover Publications, New York.

M. L. Cohen and T. K. Bergstresser (1960). *Phys. Rev.*, **141**, 787-796.

A. Einstein (1905). *Ann. Phys.*, **17**, 132.

R. H. Fowler and L. W. Nordheim (1925). *Proc. Roy. Soc.*, **A119**, 173.

R. de L. Kronig and W. G. Penney (1931). *Proc. Roy. Soc.*, **A130**, 499.

E. W. Muller and T. T. Tsong (1969). *Field Ion Microscopy Principles*, Elsevier, Amsterdam.

5

Electrons and Holes in Semiconductors

A semiconductor can generally be treated in a similar way to a free electron solid. The energies corresponding to the permitted bands are first calculated. The levels in each band are then filled with electrons until the supply has been exhausted. As previously defined, the Fermi level is the top-most filled level at zero Kelvin. As in the Free Electron model, it is only those electrons in the vicinity of the Fermi level which are significant in the conduction process. Therefore, our treatment can be restricted to a study of the properties of the highest filled and lowest empty energy bands, which are more commonly known as the *valence band* and the *conduction band*.

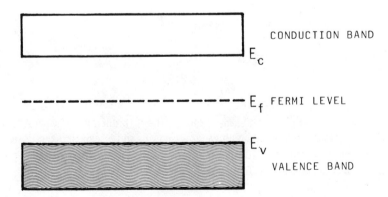

Figure 5.1 An energy band diagram for an intrinsic semiconductor.

The number of carriers which occupy levels above the Fermi level at temperatures above zero Kelvin (that is, which are in the conduction band) can be given by

$$n = S(E)\, P(E) = N_c \exp\left[\frac{-(E_c - E_f)}{kT}\right]$$

N_c is the density of states in the conduction band and is given by

$$2\left[\frac{2\pi m_e^* kT}{h^2}\right]^{3/2}$$

where m_e^* is the effective mass of electrons in that band.

If an electron is excited from the valence band, its departure leaves a hole behind it. The hole has the opposite charge to the electron but otherwise can be treated in an identical manner. Accordingly, the hole concentration is given by

$$p = N_v \exp\left[\frac{-(E_f - E_v)}{kT}\right]$$

where N_v is the density of states in the valence band and is given by

$$2\left[\frac{2\pi m_p^* kT}{h^2}\right]^{3/2}$$

m_p^* is the effective mass of holes in the valence band.

The position of the Fermi level in an intrinsic semiconductor

An intrinsic semiconductor is defined as one which contains no electrically active impurities. Thus, any electrical conduction must be due to the excitation of electrons from the valence band into the conduction band. One can therefore say that in an intrinsic semiconductor, $n = p$.

$$N_c \exp\left[\frac{-(E_c - E_f)}{kT}\right] = N_v \exp\left[\frac{-(E_f - E_v)}{kT}\right]$$

Therefore

$$E_f = \frac{E_c + E_v}{2} + \frac{kT}{2} \log_e\left[\frac{N_v}{N_c}\right]$$

The Fermi level would be at the mid-point of the forbidden gap if $N_c = N_v$. If the definitions for N_c and N_v given above are used, it can be seen that the position of the Fermi level is displaced from the mid-point by a factor which depends on the ratio of electron and hole effective masses.

$$E_f = \frac{E_c + E_v}{2} + \frac{3kT}{4} \log_e\left[\frac{m_p^*}{m_e^*}\right]$$

The position of the Fermi level in an extrinsic semiconductor

An extrinsic semiconductor is one which contains electrically active impurities. It is possible to dope a semiconductor so that $p = n$. Materials of this type are called *compensated semiconductors*. In general, however, one impurity dominates so that the semiconductor is either n-type or p-type. Figure 5.2 shows an energy band diagram at room temperature. It has a concentration N_d of donors in an energy level E_d. A Fermi–Dirac distribution is superimposed on the diagram.

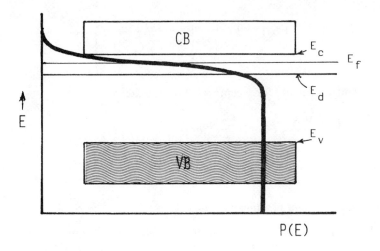

Figure 5.2 A Fermi–Dirac distribution superimposed on the energy band diagram for an extrinsic semiconductor. The donor level at E_d had a doping density N_d per unit volume.

Concentration of full donors = $N_d P(E_d)$.
Concentration of empty donors = $N_d(1 - P(E_d))$.
Every empty donor level represents one electron in the conduction band.

$$n = 2\left[\frac{2\pi m_e^* kT}{h^2}\right]^{3/2} \exp\left[\frac{-(E_c - E_f)}{kT}\right] = N_d[1 - P(E_d)]$$

$$= P(E_d) = \left[1 + \exp\left[\frac{E_d - E_f}{kT}\right]\right]^{-1}$$

Put $(E_f - E_d)/kT = x$

therefore

$$N_d[1 - (1 + \exp(-x))^{-1}] = N_d\left[1 - \left(1 - \exp(-x) + \frac{\exp(-2x)}{2} - \frac{\exp(-3x)}{6} + \ldots\right)\right]$$

if x is sufficiently large.

$$N_d[1 - (1 + \exp(-x))^{-1}] \approx N_d \exp(-x)$$

for $E_f - E_d \gg kT$

$$N_d \exp\left[\frac{-(E_f - E_d)}{kT}\right] = N_c \exp\left[\frac{-(E_c - E_f)}{kT}\right]$$

therefore

$$E_f = \frac{(E_d + E_c)}{2} + \frac{kT}{2} \log_e \left[\frac{N_d}{2\left[\frac{2\pi m_e^* kT}{h^2}\right]^{3/2}}\right]$$

Using a similar argument for acceptors in p-type material

$$E_f = \frac{(E_A + E_v)}{2} - \frac{kT}{2} \log_e \left[\frac{N_A}{2\left[\frac{2\pi m_p^* kT}{h^2}\right]}\right]$$

Carrier concentrations in semiconductors

The electron (hole) concentration in a semiconductor can be given in terms of the intrinsic concentration, n_i, and the intrinsic Fermi level, E_i, which was defined previously as

$$E_i = \frac{E_g}{2} + \frac{3kT}{4} \log_e \left[\frac{m_p^*}{m_e^*}\right] \approx \frac{E_g}{2}$$

The concentration of electrons in an n-type semiconductor, n_n, is given by

$$n_n = n_i \exp\left[\frac{E_f - E_i}{kT}\right]$$

The subscript, n, is used to indicate that the electrons are in n-type material. The concentration of holes in the same n-type material, p_n, is given by

$$p_n = n_i \exp\left[\frac{E_i - E_f}{kT}\right]$$

The product $p_n n_n$ can be seen to be equal to n_i^2.

If one uses a similar set of definitions for p-type materials, it can be shown that the product $p_p n_p = n_i^2$.

ELECTRONS AND HOLES IN SEMICONDUCTORS

Since this product, n_i^2, is independent of semiconductor type it is usual to simply write $pn = n_i^2$.

The product $pn = n_i^2$ represents a dynamic equilibrium. Electron-hole pairs are continuously being created and destroyed. It is the product of the time-averaged concentration which remains a constant. This type of phenomenon is well-known in chemistry. For water the product of the concentration of hydrogen ions (H^+) and hydroxyl ions (OH^-) = 10^{-14}. If water is pure then the concentration of hydrogen ions = 10^{-7}, the basis for pure water having a pH of 7.

The expression $pn = n_i^2$ can be used to obtain a value for n_i.

$$n_i^2 = N_c \exp\left[\frac{-(E_c - E_f)}{kT}\right] N_v \exp\left[\frac{-(E_f - E_v)}{kT}\right]$$

therefore

$$n_i = (N_c N_v)^{1/2} \exp\left[\frac{-(E_c - E_v)}{2kT}\right]$$

$$= (N_c N_v)^{1/2} \exp\left[\frac{-E_g}{2kT}\right]$$

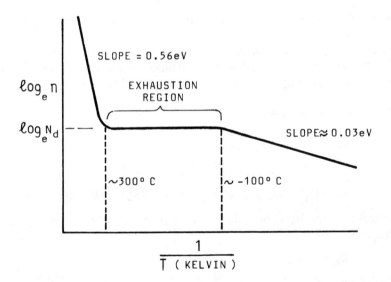

Figure 5.3 A plot of $\log_e n$ versus $1/T$ for a 0.5 Ω m n-type silicon sample.

A plot of $\log_e n_i$ versus $1/T$ in an intrinsic semiconductor gives a straight line with a slope which is half the band gap energy. The slope of a similar plot for an extrinsic semiconductor is simply the doping energy level. This can be seen experimentally in Hall effect measurements, which are an extremely effective method of characterising a semiconductor. The Hall coefficient, R_H, is equal to the reciprocal of the charge concentration ($1/(ne)$ or $1/(pe)$). The impurity type can be determined from the sign of the Hall voltage. Figure 5.3 shows a plot of $\log n$ versus $1/T$ for a sample of n-type silicon. Three distinct regions of conduction can be seen. At temperatures below about $-100°C$, carriers are being excited from the impurity level. The slope of the graph gives a value for the donor energy. The plateau is often called the 'exhaustion region'. This is where virtually all carriers have escaped from the donor impurities and are now contributing to conduction. Thus, a Hall plot can immediately yield the doping density, N_d, or acceptor density, N_A, in the case of a p-type semiconductor. Above about $300°C$, the probability of exciting electrons across the forbidden gap increases significantly and thereafter the semiconductor behaves as if it were an intrinsic material. The slope of the Hall plot in this region is half the band gap energy.

Majority and minority carriers

The concept of majority and minority carriers has already been implicitly introduced.

p_p and n_n are respectively the concentrations of majority carriers in p-type and n-type materials.

n_p and p_n are the corresponding concentrations of minority carriers.

Carrier concentrations can be determined using the principle of space charge neutrality. This states that the total sum of charges within a semiconductor must be zero.

Total charge density $\rho = e\,(p + N_d - n - N_A) = 0$
Therefore $p - n = N_A - N_d$
This can be used with $n_n p_n = n_i^2 = n_p p_p$ to give

$$n_n = \tfrac{1}{2}[(N_d - N_A) \pm \sqrt{(N_d - N_A)^2 + 4n_i^2}\,]$$
$$p_p = \tfrac{1}{2}[(N_A - N_d) \pm \sqrt{(N_A - N_d)^2 + 4n_i^2}\,]$$

Except in compensated or near-intrinsic semiconductors, where $|N_d - N_A| \approx n_i$, this reduces to

$$n_n = N_d - N_A$$
$$p_p = N_A - N_d$$

These values for majority carrier concentrations can be combined with $n_n p_n = n_i^2 = n_p p_p$ to give the minority carrier concentrations.

$$p_n = n_i^2/(N_d - N_A)$$
$$n_p = n_i^2/(N_A - N_d)$$

The motion of carriers in semiconductors

1. Diffusion

Diffusion is a well-known phenomenon in nature. Wherever there is a concentration gradient of a species, material will move down the gradient in an attempt to reduce the magnitude of the gradient to zero. If, for some reason, an accumulation of electrons occurs at one end of a semiconductor, they will then diffuse in an attempt to distribute themselves uniformly. This is expressed by Fick's first law of diffusion which states that the flux of carriers (number per second passing through unit area) is proportional to the gradient. For electrons

$$F_n = -D\,(dn/dx)$$

The negative sign indicates that electrons move down the spatial gradient (dn/dx). The constant of proportionality, D, is called the *diffusion constant*. It has dimensions of distance^2 time^{-1}.

2. Drift

If an electric field is applied to a semiconductor, the carriers will move at a velocity which is proportional to the magnitude of the field.

$$v = u\mathcal{E}$$

where $\mathcal{E}$ is the electric field. The constant of proportionality is called the *mobility*. It has dimensions distance^2 volts^{-1} time^{-1}.

Mobility and diffusion constant can be equated through the Einstein relationship

$$D = \mu\,\frac{kT}{e}$$

Carrier mobility is an extremely important parameter. It is affected by temperature, doping concentration and the magnitude of an applied electric field. It can be determined if resistivity measurements are undertaken in conjunction with Hall-temperature measurements, similar to those shown in figure 5.3.

Since resistivity $\rho = (neu)^{-1}$ and $R_H = (ne)^{-1}$ in an n-type semiconductor

$$\mu = R_H/\rho$$

Figure 5.4 shows a graph of mobility *versus* $1/T$ for a sample of silicon. At temperatures above 35 K there is a $T^{-3/2}$ relationship which is characteristic of carriers which are scattered by the low-frequency thermal vibrations in the solid. At lower temperatures, the relationship is reversed when carriers are scattered by ionised impurities. Obviously this form of scattering is strongly dependent on the doping concentration.

Carrier mobility is also dependent on the magnitude of electric field. The field dependence of electron and hole mobility in silicon is shown in figure 5.5 where it can be seen that holes are significantly less mobile than electrons. This effect,

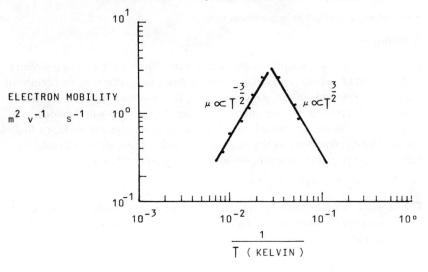

Figure 5.4 Carrier mobility *versus* $1/T$ for a 0.5 Ω m n-type silicon sample.

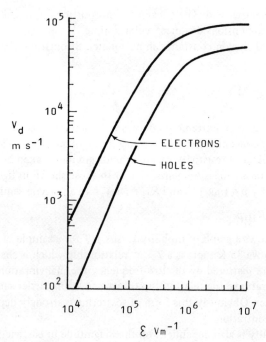

Figure 5.5 The variation of carrier drift velocity as a function of electric field for silicon.

which is due to the relative masses of the two carrier types in their respective Brillouin zones, was mentioned in the previous chapter. It will be remembered that in a reduced k-space diagram a steep parabola implies that the effective mass is small and therefore the mobility is large. At high electric fields the mobility tends to a saturation value. This occurs when the velocity of carriers starts to become comparable with the average thermal velocities inside the solid. It is obvious that Ohm's law no longer holds true at these high fields.

The relative steepness of the E-k parabolas is the controlling factor in the operation of a Gunn microwave diode. In a piece of n-type gallium arsenide, electrons are initially excited into a fairly steep-sloped direct conduction band (see figure 4.21b). However, as the field is increased, progressively more electrons acquire additional energy and transfer to the indirect conduction band where they have a lower mobility. Thus, as the electric field increases, the mobility averaged out over the entire population of carriers decreases. A decrease in mobility with increasing field is the same as a decrease in current with increasing applied bias. In this situation the material represents a negative resistance which will cause the diode to oscillate at very high frequencies.

Resistivity measurements

In theory, the resistivity of a sample can be calculated from a resistance measurement and the dimensions of the sample. However, measurements on semiconducting materials are often troubled by high-resistance contacts or by contacts which may be non-ohmic. Thus, the measured resistance is the sum of the sample resistance in series with the resistance presented by the two contacts. Contact-resistance problems can be overcome by the use of a four-contact measurement: one pair of contacts provides a source of current and the other pair monitors the sample voltage via a high input impedance voltmeter.

Sheet resistance

There are many instances, such as diffused layers, where the resistivity is a function of depth below the sample surface. A simple measurement would yield the mean resistivity which is not particularly informative. This is where the concept of sheet resistance is very useful.

In a rectangular sample the resistance is given by

$$R = \rho \frac{l}{wt}$$

where l = length, w = width and t = thickness.

A square sample would have $l = w$ and the resistance, designated by ρ_s, is independent of the magnitude of $l = w$ and is called the *sheet resistance*. Thus, a 10 μm × 10 μm sample with a resistance of 100 Ω will still represent 100 Ω even if it is enlarged to 1 mm × 1 mm.

$$\text{Sheet resistance} = \rho/t$$

It is very easy to determine the sheet resistance using a set of four in-line equally spaced sharp tip probes. There will be a very high electric field in the vicinity of a sharp tip contact and this ensures that it will display ohmic behaviour. Provided that the dimensions of a sample are very much larger than the separation between probes then

$$\rho_s = 4.532 \, V/I$$

Sheet resistance is the major parameter in the design of resistors for integrated circuits.

For example, one could consider a diffused layer on silicon which has a sheet resistance of 500 Ω per square. It is required to use it to construct a 2.5 kΩ resistor on this layer.

Since $R = \rho_s(l/w)$, regardless of the sample thickness, then a 5:1 aspect ratio will provide the required resistance even if the dimensions are 50 μm long by 10 μm wide.

Semiconductors in non-equilibrium

A semiconductor is said to be in a state of non-equilibrium if the product of concentrations pn is not equal to n_i^2. If the product is greater than n_i^2, then additional charge has been injected into the material. This can occur as a result of optical excitation with light whose energy is greater than the band gap. It can also occur as a result of heating or if the semiconductor forms part of a junction device under forward bias, as will be seen later. The product could also be less than n_i^2. This might arise as a result of carrier extraction. We are more concerned with carrier injection and with the behaviour of the semiconductor during the period when it is returning to the equilibrium state.

Carrier injection is generally divided into two classifications. It is defined as being low level when the increase in the number of carriers is significantly less than the doping density. It is much easier to treat mathematically than high-level injection, which is said to occur when the increase is of the same order as the doping density.

An example of low-level injection

If $N_d = 10^{22}$ m^{-3}, then $n_n = 10^{22}$ m^{-3} and $p_n = 2 \times 10^{10}$ m^{-3} in equilibrium.

Now supposing that 10^{14} m^{-3} electron-hole pairs are injected into the material, then the new majority and minority carrier concentrations are as follows

$$n_n = 10^{22} + 10^{14} \text{ which is virtually } 10^{22}$$

$$p_n = 10^{10} + 10^{14} \text{ which is virtually } 10^{14}$$

The majority carrier concentration is almost unchanged but the minority carrier concentration has risen significantly. The semiconductor is no longer in equilibrium as $p_n n_n = 10^{36}$.

ELECTRONS AND HOLES IN SEMICONDUCTORS

It is obvious that the minority carrier concentration is the significant parameter in determining whether a semiconductor is in equilibrium. Thus, the remainder of this chapter will be devoted to the kinetics of generation and recombination of minority carriers.

Carrier generation and recombination

If a semiconductor is illuminated with light whose energy is greater than the band gap, then there are two carrier generation mechanisms present. These are shown in figure 5.6, part a and b respectively.

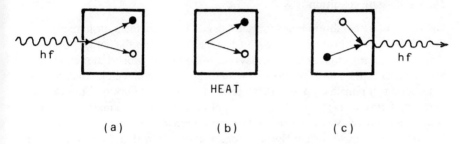

Figure 5.6 Examples of generation and recombination processes: (a) optical excitation; (b) thermal excitation; (c) spontaneous recombination.

For the purposes of the following analysis it is assumed that the rate of optical excitation is G_L, which depends on the intensity of illumination. G_{th} is the rate of thermal generation, which will depend on the temperature. Carrier recombination occurs when an electron and hole meet (figure 5.6c). The recombination rate constant, R, depends on the concentration of excess minority carriers.

The net rate of change of hole concentration in n-type material is given by

$$\frac{dp_n}{dt} = G_L + G_{th} - R = G_L - U$$

$$U = R - G_{th}$$

Where there is no illumination

$$G_L = 0 \quad \text{and} \quad \frac{dp_n}{dt} = 0$$

$U = 0$, so that $R = G_{th}$ represents a dynamic equilibrium.

U is proportional to the excess minority carrier concentration.

$$U = \frac{1}{\tau_p} [p_n - p_{n_0}]$$

p_n is the non-equilibrium minority carrier concentration. P_{n_0} is the hole concentration at equilibrium (that is, $n_n p_{n_0} = n_i^2$). τ_p is the hole minority carrier lifetime.

The rate of change of minority carriers in illuminated material is then given by

$$\frac{dp_n}{dt} = G_L - \left[\frac{p_n - p_{n_0}}{\tau_p}\right]$$

The steady-state condition

The steady-state condition is a very useful device in the mathematical manipulation of this type of problem. It is defined as the state where the net rate of change of concentration of a species is zero. If the rate at which water enters a bath is equal to the rate at which it flows out through the plug-hole, then the water level will remain constant. It is a form of pseudo-equilibrium. The concept of steady state occurs in many branches of physics. Radioactive materials such as uranium undergo successive decay through intermediates (each with their own half-life) to end up with lead. However, observations show that the abundance of the intermediates in the earth's crust does not change with time. This means that so long as there is an adequate supply of uranium, a steady-state condition exists.

In the context of this subject, a steady-state condition exists if dp_n/dt is zero. In steady state, the concentration of minority carriers is then

$$p_n = p_{n_0} + \tau_p G_L$$

Return to equilibrium

If the light is removed, the rate of generation of minority carriers, G_L, becomes zero. The carrier concentration as a function of time may then be determined from the differential equation

$$\frac{dp_n}{dt} = -\left[\frac{p_n - p_{n_0}}{\tau_p}\right]$$

In order to solve this for p_n, one can use the boundary condition that $p_n = p_L$ at $t = 0$. This gives

$$p_n(t) = p_{n_0} + (p_L - p_{n_0}) \exp[-t/\tau_p]$$

Figure 5.7 shows examples of decay of p_n as a function of time and it is obvious that the speed of restoration of equilibrium depends largely on the minority carrier lifetime.

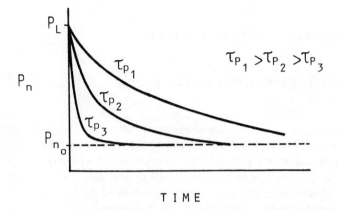

Figure 5.7 Minority carrier decay *versus* time for different values of lifetime.

Steady-state injection from a boundary

The regions of a semiconductor which are remote from the carrier injection process are not necessarily in an equilibrium state. Injection gives rise to a concentration of minority carriers which is above that of the rest of the material. Carriers will therefore diffuse along this gradient. The problem is best treated in terms of the rate of change of the excess minority carrier concentration above its equilibrium value.

$$\frac{\partial [p_n - p_{n_0}]}{\partial t} = -\frac{\partial F_p}{\partial x} - \frac{[p_n - p_{n_0}]}{\tau_p}$$

or

$$\frac{\partial [p_n - p_{n_0}]}{\partial t} = D_p \frac{\partial^2 [p_n - p_{n_0}]}{\partial x^2} - \frac{[p_n - p_{n_0}]}{\tau_p}$$

The steady-state condition can be applied to this equation to give

$$D_p \frac{\partial^2 [p_n - p_{n_0}]}{\partial x^2} = \frac{[p_n - p_{n_0}]}{\tau_p} = \frac{D_p [p_n - p_{n_0}]}{L_p^2}$$

where $L_p = \sqrt{D_p \tau_p}$ is the diffusion length.

Just as in wave mechanics, an equation of this type has solutions of the form

$$p_n - p_{n_0} = C_1 \exp [x/L_p] + C_2 \exp [-x/L_p]$$

Values for C_1 and C_2 can be obtained using the following boundary conditions

At $x = 0$, $p_n(0)$ is constant, dependent on external factors
$p_n(x)$ at infinity is equal to the equilibrium hole concentration

Therefore

$$p_n - p_{n_0} = (p_n(0) - p_{n_0}) \exp[-x/L_p]$$

or

$$p_n(x) = p_{n_0} + [p_n(0) - p_{n_0}] \exp[-x/L_p]$$

These observations can be confirmed experimentally using an arrangement similar to that used by Haynes and Schockley (1951) (see figure 5.8a). A piece of n-type silicon has small regions which are heavily doped with p-type impurity. Minority carriers are injected by applying a forward-bias pulse to the emitter. If an electric field is present, then carriers will recombine while they are moving by both diffusion and drift. Representative fractions of the remaining charge travel across the collector junctions and are displayed on an oscilloscope (see figure 5.8b).

The separation between collectors is known. Thus, the time of arrival of an injected pulse at successive collectors can be used to calculate minority carrier mobility provided that the drift field is constant.

The area under a curve is a measure of the charge at a given collector. Since the charge decays exponentially with time, a plot of $\log_e$ (area under each collector curve) *versus* time of arrival of the pulse peak can be used to calculate the minority carrier lifetime.

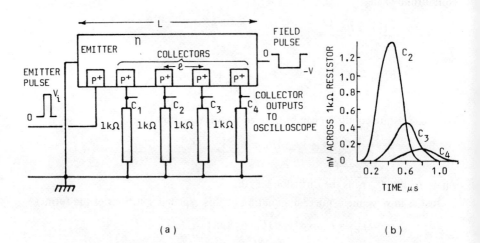

Figure 5.8 (a) Schematic arrangement for Haynes-Shockley observation of minority carrier decay based on a structure developed at the University of Birmingham. (b) Examples of the response at collectors C_2, C_3 and C_4 for a constant value of applied field.

ELECTRONS AND HOLES IN SEMICONDUCTORS

The following table gives some typical recombination parameters for silicon as a function of background doping, and it can be seen that each reduces in value as the doping is increased. This has important implications for the design and operation of many silicon semiconductor devices.

	Doping (m^{-3})		
	10^{19}	10^{20}	10^{24}
D_n ($m^2\ s^{-1}$)	4×10^{-3}	3×10^{-3}	1×10^{-3}
D_p ($m^2\ s^{-1}$)	1×10^{-3}	9×10^{-4}	5×10^{-4}
τ_n (μs)	66	30	0.5
τ_p (μs)	40	20	1.2
L_n (μm)	509	300	22
L_p (μm)	200	130	25

Problems

5.1. The effective mass of an electron in a semiconductor is $0.2m_e$. Calculate the concentration of carriers in the conduction band at:
(a) $0°C$; (b) $300°C$.
Assume that $E_c - E_f = 0.56$ eV, and is independent of temperature.

5.2. Calculate the position of the Fermi level in undoped silicon at 77 K. How is the Fermi level affected by the addition of 1 in 10^9 aluminium atoms? The band gap in silicon is 1.12 eV. The atomic weight is 28 and the density is 2329 kg m^{-3}. $m_n = 0.44m_e$ and $m_p = 0.37m_e$. The aluminium level lies 0.05 eV above the valence band.

5.3. It has been shown in the text that provided $(E_f - E_d)/kT$ is sufficiently large, then

$$n \approx N_d \exp\left[-\left(\frac{E_f - E_d}{kT}\right)\right].$$

Eliminate E_f to prove that under these conditions

$$n \approx \sqrt{N_c N_d} \exp\left[-\left(\frac{E_c - E_d}{kT}\right)\right]$$

5.4. The carrier concentration in a piece of semiconductor is measured over a range of temperatures using the Hall effect. The results show that a donor level lies 0.035 eV below the conduction band. Estimate the position of the level in relation to the valence band, given that the semiconductor becomes highly conducting at wavelengths below 9000 Å.

5.5. A piece of 0.055 Ω m p-type silicon has a carrier drift velocity of 2×10^3 m s^{-1} at an applied field of 10^5 V m^{-1}. Calculate the hole concentration and diffusivity at 27°C.

5.6. A piece of 0.5 Ω m n-type silicon has an electron mobility of 0.15 m^2 V^{-1} s^{-1}. Calculate
(a) the donor concentration
(b) the equilibrium minority carrier concentration at room temperature.

5.7. The concentration of acceptors in a bar of silicon is 5×10^{21} m^{-3}. The density of current flowing in the bar is 10^6 A m^{-2}.
(a) Calculate the average drift velocity.
(b) Estimate the volume power density given by $J^2 \rho$. (*Hint:* use the graphs in figure 5.5.)
(c) Estimate the maximum current density that can be handled before deviations in Ohm's law become obvious.

5.8. A 300 μm thick slice of germanium has an average resistivity of 0.01 Ω m. If an equi-spaced four-point probe were used to measure sheet resistance, estimate the voltage which is measured between the inner pair of probes when a current of 0.1 mA passes through the outer pair.

5.9. Holes are injected into one end of a piece of n-type silicon. The initial concentration is 10^{22} m^{-3}. If the minority carrier mobility is 0.02 m^2 V^{-1} s^{-1} and the lifetime is 10 μs, calculate the distance required for the hole concentration to drop to within 1 per cent of its equilibrium value at room temperature.

5.10. Holes are injected into the neutral n-type ($N_d = 10^{22}$ m^{-3}) side of a silicon p-n junction device at room temperature. The initial concentration of injected holes (10^{20} m^{-3}) is found to drop to 10^8 m^{-3} in 40 μm. Calculate the diffusion length for holes.

Reference

J. R. Haynes and W. Schockley (1951). *Phys. Rev.*, **81**, 835.

6

p–n Junctions

When two pieces of semiconductor of opposite types are brought together, the presence of a concentration gradient gives rise to a diffusion flux. Electrons move from n to p, leaving a region of ionised impurities of positive charge behind them. Similarly, holes move from p to n, leaving a negative space charge. Counteracting the diffusion process, there is an electrical field $\mathcal{E}$ due to charge separation. The net flux of holes is then given by the expression

$$F_p = -D_p \frac{dp}{dx} + \mu_p p \mathcal{E} \qquad (6.1)$$

Because of the positive charge on the n-type side of the junction and the negative charge on the p-type side, it is often termed the *space charge region*. It can also be seen that charge movement results in a hole concentration in the p-type side that is less than N_A and an electron concentration in the n-type side that is less than N_d. On account of this depletion of carriers, the space charge region is alternatively known as the *depletion region* or *depletion layer*.

Position of the Fermi level at equilibrium

The movement of carriers affects the relative positions of the energy levels. However, a simple analysis is possible once an equilibrium has been established. This occurs when the opposing forces of diffusion and field are equal and opposite. In this case the net flux drops to zero. The electric field is then given by the gradient in the energy levels between the p-type and n-type sides (see figure 6.1).

It is more usual to define it in terms of the gradient of the intrinsic Fermi energy E_i

$$\mathcal{E} = \frac{1}{e} \frac{dE_i}{dx} \tag{6.2}$$

Since $p = n_i \exp[(E_i - E_f/kT]$, therefore

$$\frac{dp}{dx} = \frac{p}{k}\left[\frac{dE_i}{dx} - \frac{dE_f}{dx}\right] \tag{6.3}$$

Using equations (6.2), (6.3) and the Einstein diffusivity/mobility relationship, the components of (6.1) can be replaced so that

$$F_p = \frac{D_p}{kT} p \frac{dE_f}{dx}$$

Since the hole flux at equilibrium is zero, the gradient of the Fermi level must also be zero. Therefore, one must conclude that the value of the Fermi level at equilibrium is constant across the junction. This is shown in figure 6.1.

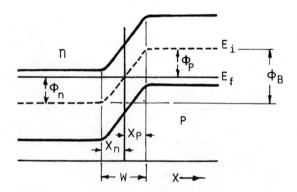

Figure 6.1 Energy band diagram for a p–n junction at equilibrium.

The amount of band bending at equilibrium

If the potential is given by ϕ, then the field

$$\mathcal{E} = \frac{1}{e} \frac{dE_i}{dx} = -\frac{d\phi}{dx}$$

In the n-type region

$$|\phi_n| = \left|\frac{E_f - E_i}{e}\right|$$

Its value can be obtained from the fact that within the depletion layer the expression for carrier density is equal to the doping density.

$$n = n_i \exp\left[\frac{(E_f - E_i)}{kT}\right] = N_d$$

so that

$$|\phi_n| = \frac{kT}{e} \log_e\left[\frac{N_d}{n_i}\right]$$

Similarly, in the p-type region

$$|\phi_p| = \left|\frac{E_i - E_f}{e}\right| = \frac{kT}{e} \log_e\left[\frac{N_A}{n_i}\right]$$

The amount of band bending $|\phi_B| = |\phi_p| + |\phi_n|$ is called the *built-in potential* of a junction.

Although the charge distribution at any point within the depletion layer is non-zero, a diode rarely develops a net positive or negative charge. This means that the total space charge within x_n must be equal and opposite to the charge within x_p. That is

$$N_d V_n = N_A V_p$$

where V_n and V_p are the volumes of the n-type and p-type space charge regions respectively.

If it can be assumed that the junction has a uniform cross-sectional area, then

$$N_d x_n = N_A x_p$$

This is an extremely useful formula which shows that the extent of depletion within one side of a junction is inversely dependent on the doping concentration.

The width of the depletion layer as a function of doping densities

The potential, ϕ, at any point within the space charge region can be determined by solving Poisson's equation

$$\frac{d^2\phi}{dx^2} = -\frac{\rho}{\epsilon_r \epsilon_0}$$

On the n-type side, Poisson's equation becomes

$$\frac{d^2\phi_n}{dx^2} = -\frac{eN_d}{\epsilon_r \epsilon_0}$$

Integrating twice gives

$$|\phi_n| = \frac{eN_d}{2\epsilon_r \epsilon_0} x_n^2$$

Similarly for the p-type side

$$|\phi_p| = \frac{eN_A}{2\epsilon_r\epsilon_0} x_p^2$$

$$\phi_B = |\phi_n| + |\phi_p| = \frac{e}{2\epsilon_r\epsilon_0}\left[N_A x_p^2 + N_d x_n^2\right]$$

$$= \frac{e}{2\epsilon_r\epsilon_0}\left[\frac{N_A^2 x_p^2}{N_A} + \frac{N_d^2 x_n^2}{N_d}\right]$$

Using $N_d x_n = N_A x_p$

$$\phi_B = \frac{e}{2\epsilon_r\epsilon_0} N_A^2 x_p^2 \left[\frac{1}{N_A} + \frac{1}{N_d}\right] = \frac{e}{2\epsilon_r\epsilon_0} N_d^2 x_n^2 \left[\frac{1}{N_A} + \frac{1}{N_d}\right]$$

so that

$$x_p = \sqrt{\frac{2\epsilon_r\epsilon_0 N_A N_d \phi_B}{eN_A^2 (N_A + N_d)}}$$

and

$$x_n = \sqrt{\frac{2\epsilon_r\epsilon_0 N_A N_d \phi_B}{eN_d^2 (N_A + N_d)}}$$

Therefore

$$x_p + x_n = \sqrt{\frac{2\epsilon_r\epsilon_0 \phi_B}{e}} \left[\sqrt{\frac{1}{\frac{N_A}{N_d}(N_A + N_d)}} + \sqrt{\frac{1}{\frac{N_d}{N_A}(N_A + N_d)}}\right]$$

$$= \sqrt{\frac{2\epsilon_r\epsilon_0 \phi_B}{e}} \left[\frac{\sqrt{\frac{N_d}{N_A}} + \sqrt{\frac{N_A}{N_d}}}{\sqrt{N_A + N_d}}\right]$$

Multiply top and bottom by $\sqrt{N_A N_d}$

$$w = x_p + x_n = \sqrt{\frac{2\epsilon_r\epsilon_0 \phi_B}{e}} \left[\frac{N_d + N_A}{\sqrt{(N_A N_d)}\sqrt{(N_A + N_d)}}\right]$$

$$= \sqrt{\frac{2\epsilon_r\epsilon_0 \phi_B}{e}} \sqrt{\frac{N_A + N_d}{N_A N_d}}$$

$$= \sqrt{\frac{2\epsilon_r\epsilon_0 \phi_B (N_A + N_d)}{e N_A N_d}}$$

In many cases the doping on one side of a junction is much greater than on the other side. Such structures are called 'single-sided junctions' on account of the way in which the depletion layer formula can be simplified.

In a p$^+$n junction $N_A \gg N_d$

$$W = \sqrt{\frac{2\epsilon_r \epsilon_0 \phi_B}{eN_d}}$$

In an n$^+$p junction $N_d \gg N_A$

$$W = \sqrt{\frac{2\epsilon_r \epsilon_0 \phi_B}{eN_A}}$$

It can be seen that, under these circumstances, the value of W is determined by the doping level on the lower doped side. The corollary of this is that only a small fraction of the entire depletion layer is distributed in the heavily doped side of the junction.

Application of bias

If an external bias is applied, the electric field within the junction changes. The carrier flux is non-zero as the diffusion and field components no longer balance. That is

$$F_p = \frac{1}{e} \mu_p p \frac{dE_f}{dx} \neq 0$$

and similarly for electrons

$$F_n = \frac{1}{e} \mu_n n \frac{dE_f}{dx} \neq 0$$

Thus E_f is no longer constant.

The difference E_f(p-type) $- E_f$(n-type) depends on the sign and magnitude of the applied bias. For reverse bias the distortion of the energy band diagram is similar to that shown in figure 6.2a. The width of the depletion layer for an n$^+$p junction is then given by

$$W = \sqrt{\frac{2\epsilon_r \epsilon_0 (\phi_B + V_R)}{eN_A}}$$

Figure 6.2b shows the band diagram for a p-n junction under forward bias. The depletion layer width decreases as a function of bias

$$W = \sqrt{\frac{2\epsilon_r \epsilon_0 (\phi_B - V_f)}{eN_A}} \quad \text{for n$^+$p}$$

W tends towards zero as the applied bias tends towards ϕ_B.

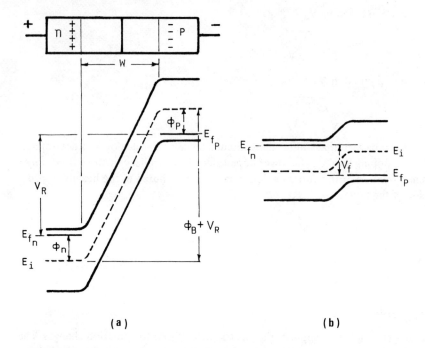

Figure 6.2 Energy band diagram for a p-n junction diode: (a) under reverse bias, (b) under forward bias.

Capacitance behaviour of p-n junctions

It was pointed out earlier that a p-n junction has a depleted region at the interface between the two types of semiconductor. This consists of an overall positive charge in the n-type side and an overall negative charge in the p-type side. The electric field inhibits carrier flow. Such a structure is not unlike the parallel plates of a capacitor, separated by a region of insulator. The value for the distance between the plates in such a capacitor is of course the depletion layer width, W.

$$C = \frac{A\epsilon_r \epsilon_0}{W} = A\sqrt{\frac{e\epsilon_r \epsilon_0 N_A}{2(\phi_B + V_R)}} \quad \text{for } n^+p$$

Since W is a function of applied bias, the structure represents a capacitance whose value depends on bias. Junction devices which are used specifically because of the variable capacitance properties are usually known as *varactor diodes*. Figure 6.3 shows a capacitance-voltage plot for a typical diode.

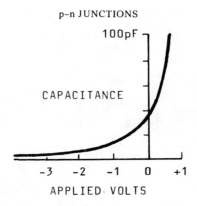

Figure 6.3 Capacitance-voltage characteristic for a p-n junction diode.

Capacitance-voltage behaviour can be used to give information about the structure of a junction device. The equation for capacitance in a p^+n diode, given above, can be written as

$$\frac{1}{C^2} = \frac{2\phi_B}{A^2 e \epsilon_r \epsilon_0 N_A} + \frac{2V_R}{A^2 e \epsilon_r \epsilon_0 N_A}$$

If C^{-2} is plotted against V (a reverse bias), then the slope of the line gives the value of doping, N_d. This value can then be used to derive ϕ_B from the intercept.

Current transport in p-n junctions

The diffusion of carriers is central to current transport in p-n junctions and will be considered first.

It has been shown that the steady-state concentration of minority carriers injected at a boundary is given by

$$p_n(x) = p_{n_0} + [p_n(0) - p_{n_0}] \exp[-x/L_p]$$

The number of carriers which surmount the junction barrier is

$$p_n(0) = p_{n_0} \exp\left[-\frac{\Delta E}{kT}\right] = p_{n_0} \exp\left[\frac{eV}{kT}\right]$$

Therefore, the number of minority carriers at point x can be written as

$$p_{n_0} + p_{n_0} \left[\exp\left(\frac{eV}{kT}\right) - 1\right] \exp\left[-\frac{x}{L_p}\right]$$

The net flux crossing the boundary is then

$$F_p(x=0) = -D_p \frac{dp_n(x=0)}{dx}$$

$$= \frac{D_p}{L_p} p_{n_0} \left[\exp\left(\frac{eV}{kT}\right) - 1\right]$$

The hole current, $eAF_p(x=0)$, is then

$$I_p = \frac{eAD_p}{L_p} p_{n_0} \left[\exp\left(\frac{eV}{kT}\right) - 1\right]$$

Reverse bias

In reverse bias

$$F_p = -D_p \frac{dp_n}{dx} + p\mu_p \mathcal{E} = 0$$

and similarly

$$F_n = -D_n \frac{dn_p}{dx} + n\mu_n \mathcal{E} = 0$$

The effect of the electric field is to increase the barrier over which thermodynamically driven carriers must pass in order to contribute to conduction. Nevertheless, the processes of thermal generation and recombination of carriers continue. Any electron-hole pairs which are formed within the sphere of influence of the field are immediately separated and contribute to conduction. The separation of charge effectively reduces the probability of recombination so that the external current is due to the generation of electron-hole pairs within the semiconductor.

There are two processes which contribute to the reverse current in a p-n diode. Carriers will form part of the conduction current if they are generated within the space charge region (figure 6.4a). This is called the *generation current*. Carriers generated in the neutral regions within one diffusion length of the depletion layer have a high probability of diffusing into that layer, where they will immediately come under the influence of the field and thereby contribute to conduction (figure 6.4b). This process gives rise to a *diffusion current*.

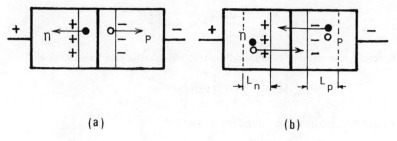

Figure 6.4 Transport processes: (a) carrier generation within the depletion layer; (b) carrier generation within a diffusion length of the depletion layer.

It can be shown that the generation current is given by

$$I = \frac{e}{2} \frac{n_i}{\tau} WA$$

where τ is the effective carrier lifetime, W is the depletion layer width and A is the junction area.

The generation current has a voltage dependence since W is a function of bias. The temperature dependence is determined by n_i. For reverse bias greater than a few kT, the hole diffusion current becomes

$$I_p = \frac{-eAD_p}{L_p} p_{n_0} = -\frac{eAD_p}{L_p} \frac{n_i^2}{N_d}$$

since $p_{n_0} N_d = n_i^2$.

The electron current is similarly

$$I_n = -\frac{eAD_n}{L_n} \frac{n_i^2}{N_A}$$

None of the terms in these expressions has a voltage dependence. The temperature dependence is determined by n_i^2. This is shown schematically in figure 6.5.

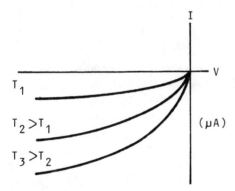

Figure 6.5 Current-voltage characteristic for a reversed p-n junction at different temperatures.

The question of which process dominates can be answered by observing the temperature and voltage dependence of reverse current–voltage characteristics. It has been found that for narrow band semiconductors, such as germanium ($E_g = 0.66$ eV), the diffusion current is more significant. Reverse conduction in silicon diodes has a much larger component of generation current. The band gap in gallium arsenide is 1.42 eV. The reverse characteristics of a diode made from this material show very little contribution from diffusion current.

Forward bias

Under forward bias

$$F_p = -D_p \frac{dp_n}{dx} + p\mu_p \mathcal{E} \neq 0$$

and similarly

$$F_n = -D_n \frac{dn_p}{dx} + n\mu_n \mathcal{E} \neq 0$$

In this case, the effect of increasing voltage is to reduce the field which inhibits carrier diffusion. Accordingly, as the field is reduced there is a net diffusion flux due to the large carrier concentration gradients. The diffusion of holes into n-type material disturbs the equilibrium, so that $p_n n_n > n_i^2$. The situation is identical for electrons injected into the p-type side of the junction. Electron–hole recombination attempts to restore the equilibrium and carriers are replaced by carriers from the contacts.

Just as in the case of reverse bias, two processes contribute to conduction. Electron–hole recombination in the space charge region gives rise to a recombination current. Electron–hole recombination in the neutral regions gives rise to a diffusion current.

It can be shown that the recombination current is given by

$$I \approx -\frac{e}{2} \frac{n_i}{\tau} WA \exp\left[\frac{eV_f}{2kT}\right]$$

This has a bias dependence through V_f and W. Since the size of the depletion region reduces under forward bias, one could assume that contributions from this source are negligible except at very low bias levels.

The diffusion current due to holes is given by

$$I = I_s \left(\exp\left[\frac{eV_f}{kT}\right] - 1\right)$$

where

$$I_s = -eAn_i^2 \left[\frac{D_n}{N_A L_n} + \frac{D_p}{N_d L_p}\right]$$

The question of which current (diffusion or recombination) dominates can be answered by observing the temperature dependence of the forward characteristics. In the case of silicon, there is a large contribution from recombination current at low forward-bias levels. Above about 0.5 V, the diffusion process dominates.

It can be seen from the above expressions that the injected current is dependent on the impurity concentration on either side of the junction. Thus, for a p^+n junction the electron current is small because N_A is large, and the hole current is large because N_d is small. One can define a junction injection efficiency, which for holes is

$$\gamma = \frac{I_p}{I_p + I_n} = \left[1 + \frac{N_d/L_n}{N_A/L_p}\right]^{-1}$$

This will be extremely important when one comes to consider bipolar junction transistors.

Junction breakdown under reverse bias

When the reverse bias on a p-n junction is increased, the field across the depletion layer rises. The ability to sustain high electric fields is limited and breakdown can occur by one of two mechanisms.

A carrier which is generated in the space charge region or which enters it from the neutral regions is accelerated in the high electric field. It gains kinetic energy ($\frac{1}{2}mv^2$). If $\frac{1}{2}mv^2 > E_g$ (the bandgap energy), then an electron-hole pair can be generated as a result of a collision between the carrier and the lattice. If within the remaining distance of the depletion layer the electron-hole pair and the original carrier can once more gain kinetic energy $> E_g$, then avalanche multiplication can occur. This is shown in figure 6.6.

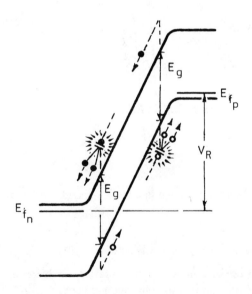

Figure 6.6 Avalanche multiplication within the depletion layer of a reverse-biased p-n junction diode.

The value of avalanche breakdown field for semiconductors such as silicon has been well characterised. This information can be used to calculate the voltage at breakdown ($V_{\text{avalanche}}$).

For an n^+p diode

$$\mathcal{E}_{\text{avalanche}} = \frac{V_{\text{avalanche}}}{W_{\text{avalanche}}} = \frac{V_{\text{avalanche}}}{\sqrt{\frac{2\epsilon_r \epsilon_0 (V_{\text{avalanche}} + \phi_B)}{eN_A}}}$$

$$= \sqrt{\frac{eN_A V_{\text{avalanche}}}{2\epsilon_r \epsilon_0}} \quad \text{if } V_{\text{avalanche}} \gg \phi_B$$

In a heavily doped p–n junction the depletion layer is very narrow. Carriers cannot gain sufficient kinetic energy within the space charge region, so that avalanche breakdown is not possible. Zener breakdown is only possible if the barrier is sufficiently narrow for quantum mechanical tunnelling to occur. This is shown in figure 6.7. For this mechanism to operate, the electric field in the depletion layer must be in excess of 10^8 V m^{-1}. This is only possible when the semiconductor doping level is 10^{24} m^{-3} or higher.

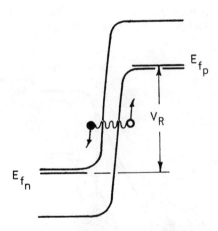

Figure 6.7 Zener breakdown in a reverse-biased p–n junction diode.

Temperature dependence of junction breakdown

When the temperature of a semiconductor is raised, the increase in lattice vibrations increases the number of collisions and hence the probability of recombination. This results in an inverse relationship between temperature and minority carrier diffusion length. A carrier moving in a high field acquires kinetic energy as it moves. If the average distance between collisions decreases, there is less likelihood of a carrier reaching an energy greater than the bandgap energy. Thus, as the temperature is increased, the reverse voltage at which avalanche breakdown can occur is also increased.

With Zener breakdown, the opposite is the case. As the temperature is increased, the energy of carriers encountering the barrier increases and so the voltage at which tunnelling occurs is reduced.

In diodes which break down in the region 4–8 V, the tunnelling and avalanche mechanisms are competing with each other. At one value (approximately 6.8 V) the dV/dT of avalanche breakdown is equal and opposite to the dV/dT of Zener breakdown. In this case, the net value of dV/dT is zero over a wide temperature range. Such devices can be used as temperature-compensated voltage reference sources.

Real diodes

The situation in real diodes is often quite different from the theory presented here. In most cases, junctions are not abrupt. For high-voltage power diodes an abrupt junction would in fact be undesirable, as it would reduce the voltage withstand capability. The doping in real junction devices is frequently achieved by diffusion, which may follow one of the well-known profiles such as Gaussian or error function complement. Alternatively, there may be a linearly graded transition from p-type to n-type.

Real diodes may be required to handle

(a) high frequencies   They must have a small junction area in order to minimise capacitance
(b) high currents   They must have good heat sinking and large area contacts
(c) high reverse voltages   See below

The problems associated with high-voltage withstand capability are perhaps the most serious. This is because our simple treatment ignores the discontinuity at the interface where semiconductor and junction end and the outside world begins. The major effect of the abrupt change is to pin the depletion layer near the surface, so that as bias is increased the field at the surface becomes considerably larger than that in the bulk. Surface breakdown well below the bulk avalanche breakdown voltage is inevitable. In the past this was overcome by making a circular device with a bevelled junction covered in silicone rubber. The rubber excluded atmospheric contamination from the junction region and ensured long-term reliability. For example, a 2000 V diode might have a bevel angle of 2° to the horizontal. It can be seen that a large area of silicon is required for maintaining voltage withstand and thus cannot carry current. Except for high-current devices this is very wasteful of silicon. Modern high-voltage diodes employ special passivating glass to cover the junction. The glass gives rise to interface charges (see chapter 8) which displace the depletion layer and reduce the surface field. This results in considerable economies, as it is possible to produce square wafers.

Many of the properties of real diodes are strongly affected by the addition of impurities such as gold or platinum which decrease the minority carrier lifetime. The minority carrier lifetime is critically important in switching diodes, although more advanced texts must be consulted for discussion of this aspect (for example, Sze (1969)).

Problems

6.1. Calculate the barrier height in a p-n junction where the doping densities in the p and n sides are 10^{24} m^{-3} and 10^{20} m^{-3} respectively.

6.2. Calculate the slope resistance of a p-n junction diode at 127°C for a forward bias of (a) 0.3 V, (b) 0.01 V. The device current at -10 V is 10^{-7} A.

6.3. The hole and electron diffusion length for the diode in problem **6.1** above are 22 μm and 100 μm respectively. Calculate the injection efficiency.

6.4. A silicon solar-cell panel consisting of an array of square diodes is to be designed using p-type islands (N_A = 10^{20} m^{-3}) diffused into an n-type (N_d = 10^{19} m^{-3}) substrate. The active area of each diode is 2.5×10^{-9} m^2 and the zero-bias capacitance is 0.03 pF. What is the maximum packing density of diodes that can be achieved if there is to be a 5 μm gap between the depletion layers of adjacent devices?

6.5. A p$^+$np$^+$ punch-through diode will only pass current when the depletion region of the reverse-biased junction touches the opposite p$^+$ layer. The voltage at which the diode starts to conduct is called the punch-through voltage. If ϕ_B = 0.6 V
   (a) At what voltage will a silicon punch-through diode start to conduct if the p$^+$ doping is 10^{24} m^{-3}, the n doping is 10^{20} m^{-3} and the separation between p$^+$ layers is 10 μm?
   (b) What is the depth of the depletion layer in the reverse-biased p$^+$ contact at punch-through?
   (c) What is the device capacitance at punch-through if the zero-bias capacitance is 100 pF?

6.6. A silicon p$^+$np$^-$p$^+$ bulk barrier diode is designed so that the x_n component of the p$^+$n junction depletion region extends throughout the entire n-type layer and touches the lightly doped p-type layer (p$^-$) at zero bias. If the p$^+$ doping is 10^{25} m^{-3} and the n doping is 10^{21} m^{-3}, calculate the dimensions of the n-layer which ensures total depletion at zero bias.

The doping of the p^- layer is 10^{20} m^{-3}. Estimate the magnitude of reverse bias which will cause the x_n component of the np^- depletion region to touch the opposing p^+ contact.

One may assume that all junctions are abrupt. $T = 25°C$.

6.7. The current in a diode due to avalanche multiplication of carriers is given by

$$I = I_0 M$$

where I_0 is the current that would pass if there were no ionisation. M is the multiplication factor, which for a silicon diode is

$$\left[1 - \left(\frac{V}{V_B}\right)^3\right]^{-1}$$

where V_B is the avalanche breakdown voltage.

If the avalanche breakdown field for silicon is 3×10^7 V m^{-1}, calculate the multiplication factor when the diode in problem **6.5** is biased at punch-through.

6.8. Two identical diodes are connected in series with a 1000 Ω resistor. 1 volt is applied to the circuit so that the diodes are both forward-biased. Calculate the power dissipated in each component at 20°C, given that the diode leakage current is 10 μA.

6.9. A power diode, 5 mm diameter, has a reverse leakage of 10^{-4} A cm^{-2} at 20°C. There is a forward voltage drop of 0.75 V when the device carries 20 A. Calculate the device series resistance.

6.10. The minority carrier lifetime, τ, in a p^+n power diode is 10 μs. It is to be used for rectifying a high-voltage, high-frequency signal, but during each positive cycle it acquires 100 coulombs m^{-3} of excess charge in the n-type layer ($N_d = 10^{20}$ m^{-3}). Until the hole concentration returns to its equilibrium value, it cannot withstand the maximum reverse voltage.

Use the simple formula $p_n(t) = p_0 \exp(-t/\tau)$ (where p_0 is the excess hole concentration at $t = 0$) to estimate the maximum frequency which can be applied to this device.

Reference

S. M. Sze (1969). *Physics of Semiconductor Devices*, Wiley, New York.

7

Junction Transistors

During the course of the previous chapter, the properties of some two terminal junction devices were examined. The text concentrated on single junctions although it is quite simple to extend the concepts to multijunction structures. The punch-through diode and bulk barrier diode (both single carrier devices) were presented for consideration as problems at the end of the chapter.

This chapter is concerned with three terminal junction structures. However, the treatment will concentrate on gate-modulated devices (transistors) and will not cover gate-controlled rectifiers such as the thyristor and triac; for these, the interested reader should consult Blicher (1976).

Some form of subdivision is necessary and it has been decided to consider single carrier (unipolar) devices first, and then to treat the bipolar transistor.

Unipolar (field effect) transistors

As mentioned in chapter 5, the resistance of a piece of material depends on the physical dimensions and its resistivity. In a semiconductor, the resistivity is determined by the doping. A junction field effect transistor (**JFET**) is a semiconductor resistor whose magnitude may be altered by means of bias applied to an additional contact called the *gate*. It consists of a layer of high resistivity material which is sandwiched by two p-n junctions. If the junctions are biased, then the dimensions of the conducting path are altered. An n-channel JFET is shown in figure 7.1. The terminals to the n-type layer are called the *source* and *drain* respectively.

The channel resistance is given by

$$R = \frac{(L_c - 2W)}{ne\mu_n(W_c - 2W)(t - 2W)}$$

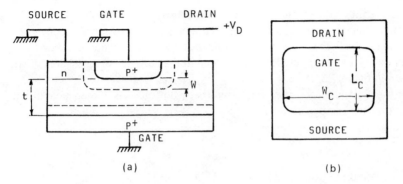

Figure 7.1 (a) Schematic cross-section of a junction field effect transistor showing the metallurgical channel thickness t and depletion width W. (b) View of a diffused JFET showing channel dimensions L_c and W_c.

where W is the depletion layer width, L_c is the channel length, W_c is the channel width and t is the separation between the two p-n junctions. In most cases, t is of the order of 5-10 μm which is of comparable magnitude to W. L_c and W_c are very much larger, so that little error is introduced if the formula is simplified to

$$R \approx \frac{L_c}{ne\mu_n W_c (t - 2W)}$$

When a bias, V_D, is applied between source and drain there will, according to Laplace's equation, be a uniform voltage drop along the length of the semiconductor. So long as the drain voltage is less than ϕ_B, W is virtually independent of the magnitude of V_D. The device then behaves as an ideal resistor. If, however, V_D is larger than ϕ_B then the magnitude of the voltage at any position, x, under the gate will affect the local depletion width $W(x)$, as shown in figure 7.2a.

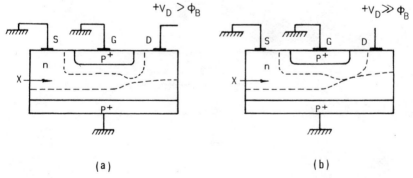

Figure 7.2 (a) Development of the space charge region below saturation. (b) The space charge region at the saturation point.

As a result of the voltage drop, W will be largest near the drain. When $V_D \gg \phi_B$ the two depletion regions near the drain touch, and the resistance becomes infinite (figure 7.2b). Thus, for large values of V_D the dependence of current, I_d, on drain voltage saturates. This can be seen in the current-voltage characteristic shown in figure 7.3.

If the metallurgical distance between the two p$^+$ gates is d, then a condition for saturation can be defined as

$$W = \frac{t}{2} = \sqrt{\frac{2\epsilon_r \epsilon_0 (V_D(\text{sat.}) + \phi_B)}{eN_d}}$$

and the voltage at saturation is

$$V_D(\text{sat.}) = \frac{eN_d t^2}{8\epsilon_r \epsilon_0} - \phi_B$$

If the gates (coupled together) are biased negatively with respect to ground potential, then saturation occurs at lower levels of V_D and accordingly less current flows. The general formula is

$$V_D(\text{sat.}) = \frac{eN_d t^2}{8\epsilon_r \epsilon_0} - \phi_B + V_g$$

where V_g is negative. A family of current-voltage characteristics are shown in figure 7.3.

The JFET is thus a unipolar (one carrier type) device whose gate can be used to modulate source-drain current flow. However, since the gate consists of reverse-biased p$^+$n junctions, the input impedance of the gate is high. By contrast, it will be seen that a bipolar junction transistor has a very low input impedance at the base, which is equivalent to the gate in the JFET.

Since the JFET is a unipolar device, the minority carrier equilibrium ($pn = n_i^2$) is only slightly perturbed during conduction. There is little or no charge storage and therefore it is suitable for use at high frequencies.

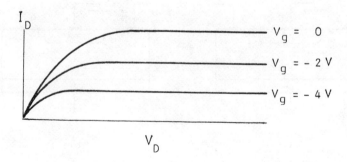

Figure 7.3 A family of current-voltage characteristics for a JFET.

Bipolar transistors

A bipolar transistor can be formed by having two p-n junctions a short distance apart within the same piece of semiconductor. Transistor action may occur whenever injected carriers at one junction contribute to current flow in the other. Figure 7.4a shows a band diagram for a typical bipolar transistor at zero bias. Figure 7.4b shows a band diagram for a transistor biased in one of its normal modes. J_1 is forward-biased and J_2 is reverse-biased. In this mode of operation a large number of holes travel from the p-type side of J_1 (the emitter) through the n-type material (the base) to the p-type side of J_2 (the collector). Thus, although J_2 is reverse-biased a large current can be introduced by external influences, provided recombination effects in the base can be minimised.

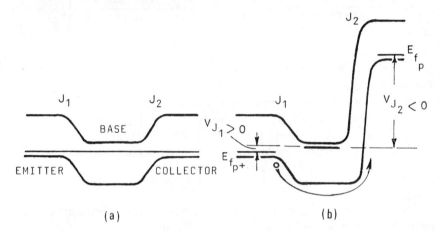

Figure 7.4 Energy band diagrams for a bipolar transistor: (a) when J_1 and J_2 are unbiased. (b) when J_1 is forward-biased and J_2 is reverse-biased.

Figure 7.5 shows a transistor in what is called a common base configuration. It shows the relative sizes of the depletion layers at the two junctions as well as the various sources of terminal current. It is quite clear that the device is indeed bipolar in its operation.

The collector terminal current, I_C, is composed of carriers injected at J_1 and carriers thermally generated in the region of J_2. This latter component represents a saturation or leakage current and is often designated I_{C_0}. Careful design of transistors ensures that it is minimised. The base current, I_B, also contains a component from electrons which are thermally generated in the region of J_2. However, the major contributions come from electrons which cross J_1 and electron-hole recombination in the base region. Electron-hole recombination becomes the dominant component if J_1 is a p^+n junction. In this case, the current is largely due to injected holes and this ensures that I_B is a small fraction of the emitter current, I_E.

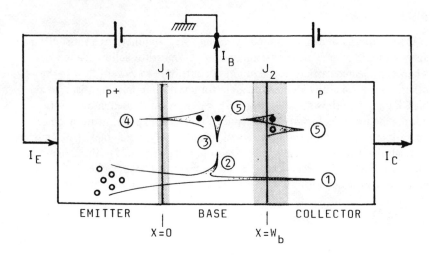

Figure 7.5 A pnp transistor in common base configuration showing the various sources of terminal current. ① Holes injected at J_1 reach J_2. ② Holes injected at J_1 recombine in base. ③ Electrons from base recombine with injected holes. ④ Electrons from base cross J_1. ⑤ Electrons and holes thermally generated in the region of J_2.

The steady-state terminal currents (I_E, I_B and I_C) within a transistor of uniform cross-section can be easily calculated if a number of simplifying assumptions are made. The assumptions, which will be modified later are:

1. The collector current is entirely due to the diffusion of holes from the emitter.
2. The electron component (source 4) of the emitter current is negligible compared with the hole current (source 1).
3. The concentration of injected holes is then

At the emitter junction

$$p_n(0) = p_{n_0} \left[\exp\left[\frac{eV_{EB}}{kT}\right] - 1 \right] = p_{n_0} \exp\left[\frac{eV_{EB}}{kT}\right]$$

if $V_{EB} \gg kT$

At the collector junction

$$p_n(W_b) = p_{n_0} \left[\exp\left[\frac{eV_{CB}}{kT}\right] - 1 \right] = -p_{n_0}$$

if $V_{CB} \ll 0$

The steady-state equation

$$\frac{\partial^2 (p_n - p_{n_0})}{\partial x^2} - \frac{(p_n - p_{n_0})}{L_p^2} = 0$$

can be solved in a similar manner to that which was used in wave mechanics.

$$p_n(x) = \alpha \exp\left[\frac{x}{L_p}\right] + \beta \exp\left[-\frac{x}{L_p}\right]$$

at $x = 0$ $p_n(0) = \alpha + \beta = p_{n_0} \exp\left[\frac{eV_{EB}}{kT}\right]$

at $x = W_b$ $p_n(W_b) = \alpha \exp\left[\frac{W_b}{L_p}\right] + \beta \exp\left[-\frac{W_b}{L_p}\right]$

Therefore

$$\alpha = \frac{p_n(W_b) - p_n(0)\exp\left[-\dfrac{W_b}{L_p}\right]}{\exp\left[\dfrac{W_b}{L_p}\right] - \exp\left[-\dfrac{W_b}{L_p}\right]}$$

$$\beta = \frac{p_n(0)\exp\left[\dfrac{W_b}{L_p}\right] - p_n(W_b)}{\exp\left[\dfrac{W_b}{L_p}\right] - \exp\left[-\dfrac{W_b}{L_p}\right]}$$

$$p_n(x) = \frac{p_n(0)\exp\left[\dfrac{W_b - x}{L_p}\right] - \exp\left[-\dfrac{(W_b - x)}{L_p}\right]}{\exp\left[\dfrac{W_b}{L_p}\right] - \exp\left[-\dfrac{W_b}{L_p}\right]}$$

since $p_n(W_b) \ll p_n(0)$

hence

$$I_p(x) = -eAD_p \frac{\partial p_n(x)}{\partial x}$$

In order to simplify the algebra, it is assumed that there is unity injection efficiency (initial assumption 2). If this is the case then the emitter current contains no electron component.

$$I_E = I_p(0) = \frac{eAD_p}{L_p}(\beta - \alpha)$$

If the collector leakage current I_{C_0} is ignored (initial assumption 1) then

$$I_C = I_p(W_b) = \frac{eAD_p}{L_p}\left[\beta\exp\left[-\frac{W_b}{L_p}\right] - \alpha\exp\left[\frac{W_b}{L_p}\right]\right]$$

After substituting for α and β one gets

$$I_E = \frac{eAD_p}{L_p}\left[p_{n_0}\cotanh\left[\frac{W_b}{L_p}\right] - p_n(W_b)\cosech\left[\frac{W_b}{L_p}\right]\right]$$

$$I_C = \frac{eAD_p}{L_p}\left[p_n(0)\cosech\left[\frac{W_b}{L_p}\right] - p_n(W_b)\cotanh\left[\frac{W_b}{L_p}\right]\right]$$

and

$$I_B = I_E - I_C$$
$$= \frac{eAD_p}{L_p}\left[p_n(W_b) + p_n(0)\tanh\left[\frac{W_b}{L_p}\right]\right]$$

Provided the emitter-base junction is forward-biased and the collector-base junction is reverse-biased, further simplifications can be made.

$$p_n(W_b) \approx -p_{n_0} \ll p_n(0)$$

$$I_E \approx \frac{eAD_p}{L_p} p_n(0)\cotanh\left[\frac{W_b}{L_p}\right]$$

but since

$$\cotanh x = \frac{1}{x} + \frac{x}{3} - \frac{x^3}{45} + \ldots$$

then provided $W_b \ll L_p$

$$I_E \approx \frac{eAD_p}{L_p} p_n(0)\left[\frac{1}{[W_b/L_p]} + \frac{[W_b/L_p]}{3}\right]$$

$$I_C \approx \frac{eAD_p}{L_p} p_n(0)\cosech\left[\frac{W_b}{L_p}\right]$$

and since

$$\cosech x = \left[\frac{1}{x} - \frac{x}{6} + \frac{7x^3}{360} \ldots\right]$$

then provided $W_b \ll L_p$

$$I_C \approx \frac{eAD_p}{L_p} p_n(0)\left[\frac{1}{[W_b/L_p]} - \frac{[W_b/L_p]}{6}\right]$$

$$I_B \approx \frac{eAD_p}{L_p} p_n(0)\tanh\left[\frac{W_b}{2L_p}\right]$$

and since

$$\left[\tanh x = x - \frac{x^3}{3} \cdots \right]$$

then provided $W_b \ll L_p$

$$I_B \approx \frac{eAD_p}{L_p} p_n(0) \frac{W_b}{2L_p} = \frac{eAW_b}{2\tau_p} p_n(0)$$

Transistor gain

It has been mentioned previously that the emitter current is equal to the sum of the base and collector currents, and that for good transistor design one attempts to arrange that I_B is only a very small fraction of I_E. The circuit designer who employs bipolar transistors uses several parameters to define performance. One of these is the common base gain α.

$$\alpha = I_C/I_E = \gamma B$$

where γ is the emitter injection efficiency and B is the base transport factor. γ is defined as the ratio of the emitter hole current to the total emitter current.

$$\gamma = \frac{I_{E_p}}{I_{E_p} + I_{E_n}} \approx 1 \quad \text{if } I_{E_n} \ll I_{E_p}$$

B is the ratio of the total collector current to the emitter hole current

$$B = \frac{I_C}{I_{E_p}}$$

It is the fraction of holes which cross the base without recombining.

If the base width is very small then $\alpha \approx 1$.

The common emitter gain, β, is another design parameter. It is defined as the ratio of the collector current to the base current.

$$\beta = I_C/I_B = \alpha/(1 - \alpha)$$

β is normally very much greater than unity.

Non-ideal injection efficiency

The derivations of the terminal currents were made on the assumption that the injection efficiency, γ, was very close to unity. It simplified the algebra. This assumption is not always valid and the factors which affect γ and B should now be estimated.

$$\gamma = \frac{I_{E_p}}{I_{E_p} + I_{E_n}} = \left[1 + \frac{I_{E_n}}{I_{E_p}}\right]^{-1}$$

It can be shown from p-n junction theory that

$$\gamma = \left[1 + \frac{\left[\dfrac{D_n N_d}{L_n}\right]}{\left[\dfrac{D_p N_A}{L_p}\right]} \tanh\left[\frac{W_b}{L_p}\right]\right]^{-1}$$

and, provided the length of the emitter is large

$$\gamma = \left[1 + \frac{\sigma_n L_p}{\sigma_p L_n}\frac{W_b}{L_p}\right]^{-1} = \left[1 + \frac{\sigma_n W_b}{\sigma_p L_n}\right]^{-1}$$

Thus, in a pnp transistor the factor $\sigma_n W_b / \sigma_p L_n$ should be kept as small as possible in order to obtain an injection efficiency close to unity.

It can also be shown for the base transport factor that

$$B = \frac{I_C}{I_{E_p}} \approx \mathrm{sech}\left(\frac{W_b}{L_p}\right) = 1 - \left[\frac{W_b}{2L_p}\right]^2$$

Once again this confirms the importance of having a very narrow base.

Some secondary effects on bipolar transistor performance

1. Base region drift

It will be remembered from simple injection theory that an electron current is the sum of a diffusion and drift term

$$I_n = N_d e A \mu_n \mathscr{E} + e A D_n \frac{dN_d(x)}{dx}$$

The transistor derivations up to now have been for a condition where the doping in the base is uniform. This means that at equilibrium — that is, when I_n is zero — the field term must be zero since the concentration gradient is zero. Many modern transistors, however, are made using diffusion processes which introduce non-zero concentration gradients. Thus at equilibrium, in base

$$I_n(x) = N_d(x) e A \mu_n \mathscr{E}(x) + e A D_n \frac{dN_d(x)}{dx} = 0$$

therefore

$$\mathscr{E}(x) = -\frac{kT}{e}\frac{1}{N_d(x)}\frac{dN_d(x)}{dx}$$

The field is positive from emitter to collector and thus aids the motion of holes. As this decreases the time which holes spend in the base, it both reduces recombination and improves the high-frequency performance.

2. Base narrowing or Early effect

The derivations so far have assumed a base width, W_b, which is independent of external bias. The depletion layer of the forward-biased emitter junction, J_1, has little effect on W_b. However, it should be remembered that the collector junction J_2 is reverse-biased and that a fraction, x_n, of depletion layer lies within the base. Thus, as was first pointed out by J. M. Early, the base width is effectively $W_b - x_n$, where x_n is a function of collector bias. The base width and therefore the performance of the transistor is modulated by the magnitude of the base-collector bias. Ultimately, if $x_n(V_{CB}) = W_b$, then the depletion layer has punched through and the current flow within the device is no longer influenced by base current.

Some considerations for good bipolar transistor design

1. The first and most important consideration is, of course, that the base width should be as narrow as reasonably possible. The minimum width will be subject to limitations imposed by punch-through breakdown. This is dependent on relative doping levels.
2. The next is that the emitter should be heavily doped compared with the base. This ensures that the injection efficiency is close to unity. Once more there are trade-offs. Injection efficiency could be optimised by having a low base doping. However, the component of the base–collector depletion layer within the base increases as the base doping is reduced in relation to the collector doping. This means that if punch-through breakdown is to be avoided, the base doping must be significantly higher than the collector doping.
3. The base transport factor is kept close to unity by having a narrow base which is lightly doped. Again, there are inter-related trade-offs. As the emitter doping is increased, L_p becomes smaller and therefore the $W_b/2L_p$ term has a stronger influence on the value of B.

Thus, as in most aspects of semiconductor devices, the various parameters which affect a bipolar junction transistor must be balanced relative to each other in order to approach an optimum design.

Real bipolar transistors

Real transistors, like diodes, often diverge from the simplified theory presented here. Some of the earliest devices were manufactured by first thinning an area of silicon, intended as the base. The emitter and collector were formed by alloying a metal into the base region. Control of the processes was limited and it was not unusual for transistors to be individually graded into RF and AF types after they had been tested for frequency response.

118   SOLID STATE DEVICES – A QUANTUM PHYSICS APPROACH

The development of the planar process, which is central to the fabrication of integrated circuits, brought about a revolution in the design and manufacture of transistors. Nevertheless, it has brought its own inherent problems. The aspect ratio of a planar transistor is such that the surface dimensions of a device are normally much larger than the junction depths. This gives rise to collector and base access resistance problems. The collector access resistance degrades transistor performance. Base access resistance gives rise to current crowding which may cause dissipation problems, particularly in power transistors. Both can be minimised by careful design, the inclusion of subsurface heavily doped layers and inter-digitated emitter-base junction geometries.

Problems

7.1. A JFET is made by diffusing p^+ gates into 0.02 Ω m n-type silicon (2×10^{21} m^{-3}). It has the following dimensions: the gate length is 50 μm and the gate width is 40 μm. The zero-bias channel width is 5 μm. What bias must be applied to the gate to give a channel resistance of 10 Ω? [$\phi_B = 0.6$ V.]

7.2. A hypothetical junction FET is constructed by diffusing heavily doped p-type gates into opposite sides of a 10 μm thick layer of n-type silicon ($N_d = 10^{21}$ m^{-3}). At a gate bias of -10 V the device is in saturation at $V_D = 0$ and a gate capacitance of 20 pF is measured. Calculate the area and depth of the gate diffusions.

7.3. In a JFET the small element of channel length dl contributes a resistance dR given by

$$dR = \frac{dl}{W_C e \mu_n N_d (t - 2W)}$$

where W_C is the channel width, t is the channel thickness and W is the depletion width, given by

$$W = \sqrt{\frac{2\epsilon_r \epsilon_0 (V_D + \phi_B - V_G)}{eN_d}}$$

Show that an integration of $dV = I_D dR$ from $l = 0$ to $l = l_c$ gives

$$I_D = G\left[V_D - \frac{2}{3}\sqrt{\frac{8\epsilon_r \epsilon_0}{eN_d t^2}}\left[(V_D + \phi_B - V_G)^{3/2} - (\phi_B - V_G)^{3/2}\right]\right]$$

where

$$G = \frac{W_C e \mu_n N_d t}{L_c}$$

The linear region of the current-voltage characteristic corresponds to the condition $V_D \ll \phi_B - V_G$. Show that for the linear region the current in a JFET can be approximated by

$$I_D = G\left[1 - \sqrt{\frac{8\epsilon_r\epsilon_0(\phi_B - V_G)}{eN_d t^2}}\right]V_D$$

7.4. The transconductance, g_m, in a JFET is defined as the rate of change of I_D with V_G when V_D is held constant. Differentiate the general expression for I_D (given in the previous problem) to obtain a value for g_m.

Show that in the linear region of device operation

$$g_m = G\sqrt{\frac{8\epsilon_r\epsilon_0}{eN_d t^2}} \frac{V_D}{2\sqrt{\phi_B - V_G}}$$

An expression for g_m in the saturation region can be obtained by inserting the value for $V_D = V_D(\text{sat.})$. Show that

$$g_m(\text{sat.}) = G\left[1 - \sqrt{\frac{8\epsilon_r\epsilon_0(\phi_B - V_G)}{eN_d t^2}}\right]$$

7.5. A pnp transistor with an injection efficiency of 0.97 has a common base gain of 0.95. Calculate the injected hole current that gives a collector current of 10 mA.

7.6. An npn transistor has $\beta = 1000$. The base width is 0.5 μm and the electron and hole diffusion lengths are 50 μm and 20 μm respectively. Calculate the ratio of base to emitter resistivity.

7.7. An npn transistor has the following parameters: the base width is 1 μm, the base doping is 5×10^{22} m^{-3}, the emitter doping is 10^{25} m^{-3} and it can be considered to be large. Calculate
(a) β
(b) the electron charge in the base, given by $|I_B \tau_n|$ at a collector current of 100 mA.
$\mu_p = 0.065$ m^2 V^{-1} s^{-1}, $\mu_n = 0.13$ m^2 V^{-1} s^{-1}, $L_n = 80$ μm, $L_p = 10$ μm and $T = 27°C$.

7.8. A symmetrical n$^+$pn$^+$ transistor has a 2 μm base width. The base doping is 10^{22} m^{-3} and the n-doping is greater by a factor of 100. The minority carrier lifetime is 0.5 μs and the electron diffusion constant is 10^{-3} m^2 s^{-1}. Calculate the emitter injection efficiency, the base transport factor and estimate the true width of the base at a collector base bias of −10 V. It may be assumed that $\phi_B = 0.6$ V, $L_p = 80$ μm and that the emitter base junction is heavily forward-biased.

Reference

A. Blicher (1976). *Thyristor Physics*, Springer, New York.

8

Surface Effects and Surface Devices

The preceding chapters have concentrated on the structure and properties of several semiconductor devices. Up to now, contacts to the outside world have at all times been considered as ideal; that is, they do not in any way affect the operation of the device. This is normally not the case. Unless special precautions are taken, the metal contact to the semiconductor can dominate the electrical properties. This chapter covers some of the properties of semiconductor surfaces and considers how they may be utilised effectively.

The interposition of an insulating layer between metal and semiconductor inhibits DC conduction. The properties of the resulting capacitor-like structure then depend on the metal, the semiconductor and the nature and thickness of the insulator. Metal–oxide–semiconductor (MOS) structures are a subset of this class and have assumed immense importance in high-density integrated circuits. The properties of MOS interfaces and their implementation as MOS capacitors, transistors and resistors are presented. The chapter closes with some indication of the simplicity of forming planar MOS circuits.

The metal–semiconductor contact

Band structures for a separated metal and semiconductor are shown in figure 8.1. In both cases, only the important energy levels E_c, E_v as well as E_{f_s} and E_{f_m} (the Fermi level in the semiconductor and metal respectively) are shown. The quantity χ is called the electron affinity and is the work involved in taking an electron from the bottom of the conduction band out to infinity.

It can be seen that the positions of the Fermi levels are not identical with respect to the vacuum level. It is known that if the metal and semiconductor are

SURFACE EFFECTS AND SURFACE DEVICES

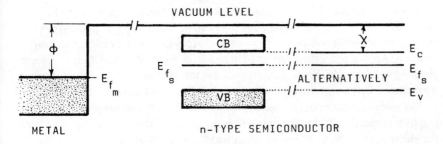

Figure 8.1 Energy band diagrams for a metal and an n-type semiconductor prior to the formation of a metal-semiconductor contact.

in contact and at equilibrium, then the Fermi levels must coincide. The relative positions of the Fermi levels in the separated metal and n-type semiconductor, shown in figure 8.1, indicate that there will be a concentration gradient if they are brought together. Thus, when a contact is made between the metal and the semiconductor, electrons will flow from the semiconductor to the metal until equilibrium is established. Therefore, the semiconductor nearest the metal will be depleted of electrons. The bands will bend so that equilibrium is reached when the diffusion force and the electrostatic force balance each other, just as in a p-n diode. This is shown in figure 8.2. The difference between work function, ϕ, and electron affinity, χ, represents a barrier of height ϕ_B over which further electrons will have to travel if the contact is to conduct (figure 8.2a).

If the metal is biased negatively with respect to the semiconductor, then the depletion width broadens as in a p-n diode (figure 8.2b). A metal-semiconductor contact which acts in this way displays non-ohmic current-voltage characteristics.

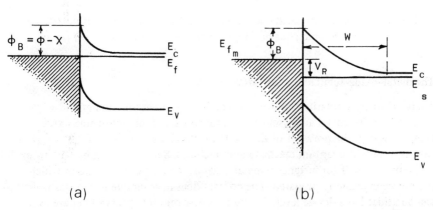

Figure 8.2 Energy band diagram for a metal-semiconductor contact: (a) at zero bias, (b) in reverse bias.

Metal-semiconductor diodes are sometimes called *Schottky diodes*. They have many advantages over p-n diodes. ϕ_B(m-s) can be much smaller than ϕ_B(p-n). Conduction involves one carrier type only; there is no minority carrier recombination. Accordingly, they can be switched from forward conduction to reverse blocking very quickly. Small Schottky diodes are used as detectors at microwave frequencies. Large-area diodes are used in switched mode power supply units.

The metal-semiconductor contact which was discussed in the previous paragraphs was specifically for the case where ϕ, the work function, was greater than the electron affinity, χ. If, on the other hand, ϕ was not greater than χ, there would be no barrier to electrons when a metal and semiconductor were brought together. Such a contact would display ohmic current-voltage characteristics.

Unfortunately, this situation is rarely realised in practice. In most cases $\phi > \chi$. The alternative method of obtaining an ohmic contact makes use of quantum mechanical tunnelling. In a heavily doped semiconductor the depletion layer will be very narrow. If it is sufficiently narrow for charge to tunnel between metal and semiconductor (figure 8.3), then the contact will appear to be ohmic. This technique is extensively used in real devices, both discrete and integrated.

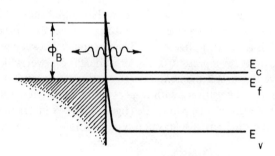

Figure 8.3 Quantum mechanical tunnelling through the contact between a metal and a heavily doped n-type semiconductor.

The metal-oxide-semiconductor (MOS) contact

There are many similarities between a metal-semiconductor contact and a contact between metal and a semiconductor which has an intervening oxide layer. However, the presence of the oxide blocks charge transport (unless it is so thin as to permit quantum mechanical tunnelling). Such a contact may be formed unintentionally, if for instance a metal is deposited on a semiconductor which has not been properly prepared. The contact does have unique properties which can be utilised at a device level. To the outside world it appears to behave as if there were two capacitors in series. The metal separated from the semiconductor by the oxide layer represents one of the capacitors. The other arises from the semiconductor surface depletion layer capacitance, owing to band bending as in a p-n junction. The bending of the bands in the semiconductor depends on the semiconductor type and the relative values of ϕ and χ.

It will be seen that the properties of the MOS contact are dominated by the semiconductor surface capacitance. If bias is applied to the metal then the semiconductor capacitance will change according to the sense and magnitude of the bias. For the purposes of the following discussion, the behaviour of a metal-oxide contact to a p-type semiconductor will be considered.

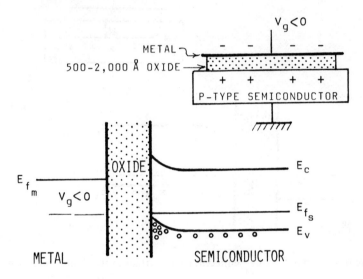

Figure 8.4 A metal-oxide-semiconductor contact in the accumulation mode.

Case 1. Metal with negative bias

A negative charge on the metal will attract additional holes to the semiconductor surface. This process is called *accumulation*. The energy bands near the metal are bent so that $E_{fs} - E_v$ at the semiconductor surface is less than $E_{fs} - E_v$ in the semiconductor bulk. Accordingly, the concentration of carriers is larger at the surface (see figure 8.4). It can be seen that for any negative value of bias to the metal there will be a positive charge separated from a negative charge by a dielectric, namely the oxide. Thus, the capacitance will be independent of the magnitude of the bias.

Case 2. Metal with slightly positive bias

Positive charge on the metal will repel holes from the semiconductor surface. The energy bands near the metal are bent so that $E_{fs} - E_v$ at the semiconductor surface is slightly greater than $E_{fs} - E_v$ in the semiconductor bulk. From the band diagram (figure 8.5) it can be seen that the concentration of holes is less at the surface. This process is called, aptly, *depletion*. There is a depletion layer similar to a p-n junction and its width is a function of bias magnitude. The capacitance of an MOS contact operating under these conditions depends on bias in an analogous way to the capacitance in a junction diode.

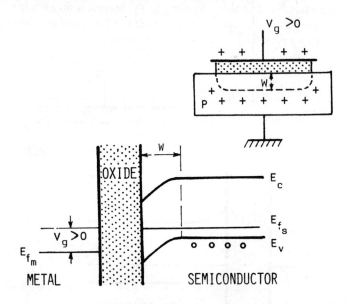

Figure 8.5 A metal-oxide-semiconductor contact in the depletion mode.

Case 3. Metal with a strong positive bias

Positive charge on the metal will strongly repel holes from the semiconductor surface. The energy bands near the metal are bent so that $E_{fs} - E_v$ at the semiconductor surface is very much greater than $E_{fs} - E_v$ in the semiconductor bulk. There comes a point (figure 8.6) when $E_c - E_{fs}$ is much less than $E_{fs} - E_v$. In band terms this means that the surface should be n-type and not p-type. In effect, this means that the minority carriers (electrons), whose bulk semiconductor concentration is given by $p_p n_p = n_i^2$, will be attracted to the oxide-semiconductor interface. Once the surface starts to become n-type the depletion layer ceases to grow further and any increase in bias is matched by an increase in the surface concentration of minority carriers. This process is called *inversion*. It is a non-equilibrium situation and if the bias is removed the surface minority carrier concentration will return to the bulk semiconductor value by recombination. It is obvious from what has been said previously that the steady-state concentration of electrons at an inverted surface is strongly dependent on the minority carrier lifetime.

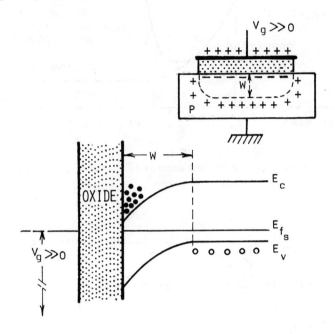

Figure 8.6 A metal-oxide-semiconductor contact in the inversion mode.

At first sight it might be assumed that once the surface is inverted, the capacitance of the MOS contact might be identical to the accumulation mode capacitance. However, this is only true if capacitance measurements are performed at very low frequencies where the recombination generation rates of minority carriers can keep up with the measuring signal. This is because it is only at low frequencies that charge exchange with the inversion layer can keep up with the changes in the applied signal. The variation of C/C_{ox} as a function of voltage is shown in figure 8.7. C_{ox} is the capacitance which would result if that thickness of oxide were sandwiched between two metal plates.

The value of C/C_{ox} during depletion can be calculated as follows. The gate voltage is distributed across the oxide and semiconductor

$$V_G = V_{ox} + \phi_s$$

where ϕ_s is the semiconductor potential.

The fields across the oxide and semiconductor are

$$\mathcal{E}_{ox} = \frac{V_{ox}}{x_{ox}} \quad (x_{ox} = \text{oxide thickness})$$

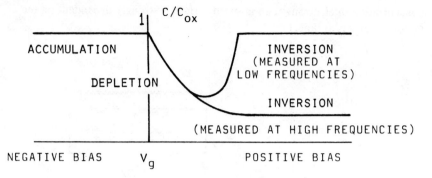

Figure 8.7 Capacitance-voltage characteristics for a metal-oxide, p-type semi-conductor contact.

$$\mathscr{E}_s = \frac{-Q_s}{\epsilon_s \epsilon_0}$$ (ϵ_s is the semiconductor permittivity, $-Q_s$ is the induced semiconductor interface charge)

The electric displacement D is continuous at the interface

$$\epsilon_{ox} \mathscr{E}_{ox} = \epsilon_s \mathscr{E}_s$$

Therefore

$$V_{ox} = \mathscr{E}_{ox} x_{ox} = \frac{\epsilon_s \mathscr{E}_s}{\epsilon_{ox}} x_{ox} = \frac{-Q_s}{\epsilon_s \epsilon_0} \frac{\epsilon_s}{\epsilon_{ox}} x_{ox}$$

$$= \frac{-Q_s}{\left[\dfrac{\epsilon_0 \epsilon_{ox}}{x_{ox}}\right]} = \frac{-Q_s}{C_{ox}}$$

thus

$$V_G = -\frac{Q_s}{C_{ox}} + \phi_s$$

If the gate voltage V_G is altered by a quantity dV_G then

$$C = \frac{dQ_G}{dV_G} = -\frac{dQ_s}{dV_G} = \frac{-dQ}{-\left[\dfrac{dQ_s}{C_{ox}}\right] + d\phi_s}$$

$$= \frac{1}{\dfrac{1}{C_{ox}} + \dfrac{1}{C_s}} \quad \text{where } C_s = \frac{dQ_s}{d\phi_s} = \frac{\epsilon_s \epsilon_0}{x_d}$$

Using the formula for a single-sided abrupt junction, x_d can be eliminated

$$x_d = \left[\frac{2\epsilon_s\epsilon_0}{eN_A} V_G\right]^{1/2}$$

$$C_s = \frac{\epsilon_s\epsilon_0}{\left[\frac{2\epsilon_s\epsilon_0 V_G}{eN_A}\right]^{1/2}} = \left[\frac{\epsilon_s\epsilon_0 eN_A}{2V_G}\right]^{1/2}$$

therefore

$$\frac{C}{C_{ox}} = \frac{1}{1 + \left[\frac{2\epsilon_0\epsilon_{ox}^2 V_G}{eN_A\epsilon_s x_{ox}^2}\right]^{1/2}}$$

The capacitance–voltage behaviour of an MOS structure is often more complicated than is predicted by the simple theory which has been outlined here. The presence of fixed interface charge displaces the capacitance–voltage spectrum to left or right, depending on the sign of the charge. For example, if the interface between the p-type semiconductor and the oxide had a positive charge which had been induced during oxide growth, then the semiconductor would be in depletion at zero bias. A negative bias would be required to overcome this charge.

Charges in oxides can also be mobile. Impurities like sodium in an oxide can result in capacitance–voltage characteristics which change with bias, due to migration of the charge in the electric field. This is one of the reasons why it is important in silicon MOS device fabrication not to touch anything with bare hands. There is sufficient sodium in the grease on one's hands for a complete processing furnace to be contaminated if one silicon slice were touched once.

Contaminants such as sodium and gold also reduce the minority carrier lifetime in the semiconductor. This can be easily observed in capacitance–voltage characteristics. As the minority carrier lifetime is reduced, the voltage necessary for inversion, the inversion threshold, is increased.

Metal-oxide-semiconductor (MOS) devices

The special properties of the MOS interface can be used in the realisation of various devices. These can be constructed as either discrete components, but largely on account of their complementary nature they are very frequently included together in MOS integrated circuits.

MOS capacitors

A fixed-value capacitor can be made by biasing the metal so that the structure operates in the accumulation mode. By operating in the depletion mode a variable capacitance is possible. This has several advantages over a conventional p-n varactor diode. The conventional varactor has a variable capacitance only in reverse bias. In forward bias the capacitance rises very quickly and the diode becomes heavily conducting. An MOS capacitance can be made variable for either positive or negative bias, depending on the doping type of the semiconductor. If, by chance, the bias applied to a depletion mode capacitor does change sign, conduction does not occur and the capacitance remains constant at its zero bias (accumulation mode) value.

MOS transistors

The MOS structure can be used in three-terminal devices where the gate voltage modulates current flow. MOS transistors can be operated using accumulation, depletion or inversion modes. The inversion mode transistor (shown in figure 8.8a) will be considered first.

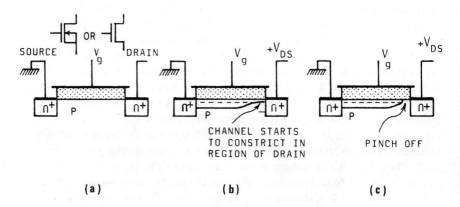

Figure 8.8 An inversion mode MOS transistor: (a) at zero bias — note symbol; (b) at $V_g > V_T$ and V_{DS} just below pinch-off; (c) at $V_g > V_T$ and V_{DS} at pinch-off.

If the gate voltage is less than the inversion threshold, V_T, then the source and drain are isolated because one junction will always be reversed relative to the other. When the gate is greater than V_T, the source and drain are connected by an inversion channel. The conductance of the channel will depend on the volume of charge which is induced in the channel and hence on the magnitude of the gate voltage. For a given V_g the drain-source current, I_{DS}, will be proportional to the drain-source voltage, V_{DS}, up to a point called pinch-off. As I_{DS} increases, the voltage distribution within the inversion layer between source and drain will be such that the potential difference between the gate and channel will be reduced. Thus, Q will decrease and the resistance of the channel will also increase (figure 8.8b). There comes a point where the voltage drop across the oxide falls below V_T, the threshold in the vicinity of the drain (figure 8.8c). This is the pinch-off point where the current–voltage characteristics (figure 8.9) become saturated.

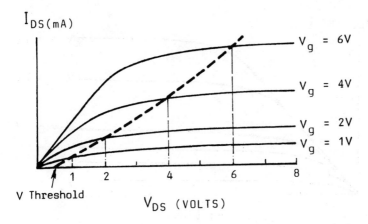

Figure 8.9 A current-voltage characteristic for an inversion mode MOS transistor with different values of V_g. The broken line represents the locus $V_{DS} = V_g$.

The inversion mode MOSFET is a normally OFF transistor. MOSFETs which operate in either the accumulation or depletion mode are normally ON at zero gate bias. The symbol for a depletion mode MOSFET is shown in figure 8.10a. The region between source and drain is merely a semiconductor resistor (figure 8.10b). The effect of an applied gate voltage either increases or decreases the

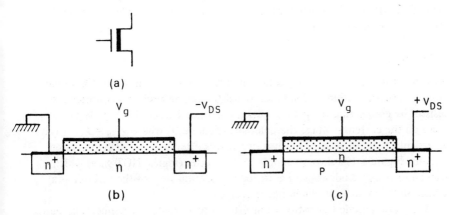

Figure 8.10 (a) Symbol for a depletion mode MOSFET. (b) A depletion mode MOSFET formed on an n-type substrate. (c) A depletion mode MOSFET formed by the placement of an n-type channel into a p-type substrate.

magnitude of the resistance. The current-voltage characteristics of an n-channel accumulation/depletion transistor are shown in figure 8.11.

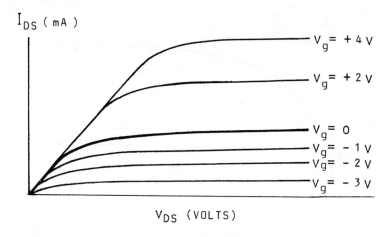

Figure 8.11 A family of current-voltage characteristics for the structure shown in figure 8.10b.

The structure shown in figure 8.10c is particularly useful since it can be easily made using n-channel inversion MOSFET (n-MOS) integrated circuit technology. A shallow n-layer is introduced between source and drain prior to the placement of the gate oxide. The only way in which the current-voltage characteristics will differ from the standard structure (figure 8.10b) is that in depletion mode there will be a cut-off gate voltage when the thin n-layer is depleted of carriers.

MOS resistors

MOS resistors can be made to operate in either the inversion or the accumulation/depletion modes. An inversion mode transistor can be easily transformed into a resistor by connecting the gate and drain together. This connection defines a locus on the current-voltage characteristic (figure 8.9) where $V_{DS} = V_g$. It is not a particularly linear resistor but it is one that is often used in MOS integrated circuits. Figure 8.12a shows a simple MOS inverter (logical NOT gate) which uses an inversion mode resistor as its load. A section view of the inverter as a planar MOS circuit is shown in figure 8.12b.

A depletion mode transistor is converted into a resistor by connecting the gate and the source together. Its characteristic is defined by the $V_g = 0$ curve in figure 8.11. Figure 8.13a shows an inverter with a depletion mode load resistor. A sectional view of the realisation as a planar MOS circuit is shown in figure 8.13b. The depletion mode resistor, like the inversion mode resistor, is not a linear device. The relationship between slope resistance and current means that at low-current levels a depletion mode resistor will have a low value. The situation with inversion mode resistors is quite the reverse. The magnitude of the resistance rises as the current through the device is reduced. This behaviour is undesirable in many digital MOS circuits, as it introduces distortion. For this reason, the depletion mode resistor is preferred even though its implementation generally involves additional processing steps.

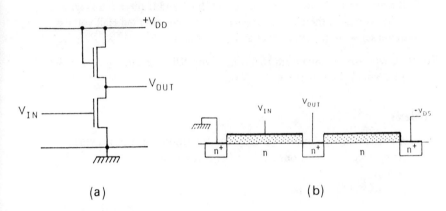

Figure 8.12 (a) An inverter circuit which uses an inversion mode resistor as a load. (b) Sectional view of the planar realisation of the circuit in part (a).

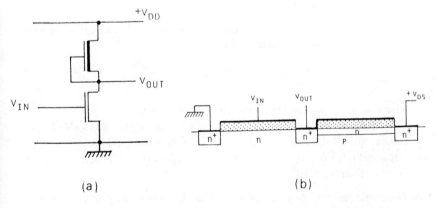

Figure 8.13 (a) An inverter circuit which uses a depletion mode resistor as a load. (b) Sectional view of the planar realisation of the circuit in part (a).

Conclusion

Quantum mechanics has been used to increase our knowledge of semiconducting materials and it has therefore been possible to develop useful devices. The subject has now come full circle. Solid state devices are being used to increase our understanding of quantum mechanics. The region immediately below the oxide in an inversion mode MOSFET is an extremely close approximation to a two-dimensional electron distribution and permits the study of two-dimensional quantum phenomena (Pepper (1986)).

If a current flows between source and drain and if there is a magnetic field normal to the plane of the gate, then the observed Hall voltage increases with quantum steps as V_g is increased.

For their work on the quantum Hall effect, von Klitzing and co-workers were awarded a Nobel prize in 1985.

Problems

8.1. A coalescence of junction theory and thermionic theory provides the current-voltage relationship for a Schottky diode.

$$J = A^*T^2 \exp\left[-\frac{e\phi_B}{kT}\right]\left(\exp\left[\frac{eV}{kT}\right] - 1\right)$$

where A^* is the effective Richardson constant = $(m^*/m_e)A$, and ϕ_B is the Schottky barrier height.

The barrier height for the aluminium/n-type silicon contact is 0.7 eV. $A^* = 0.2\ (120.0)$ amp m^{-2} Kelvin. Calculate the specific slope resistance (Ω m^2) at 25°C for bias levels of 0.1 V and -10 V. Repeat the calculation for the same bias levels at 125°C.

8.2. The extent of depletion layer in the semiconductor component of a metal/n-type semiconductor contact is given by

$$\sqrt{\frac{2\epsilon_r\epsilon_0(\phi_B + V)}{eN_d}}$$

where ϕ_B is the metal/semiconductor barrier height.

A titanium/n-type silicon contact has a barrier height of 0.5 eV. A comparison of the zero bias depletion layer widths should indicate that a semiconductor doping level of 10^{26} m^{-3} might provide a more ohmic contact than $N_d = 10^{20}$ m^{-3}.

8.3. The work function of gold is 4.8 eV. The electron affinity of 0.01 Ω m p-type silicon is 4.2 eV. Calculate the width of the depletion layer at a reverse bias of 2 V. $\mu_p = 0.038$ m^2 V^{-1} s^{-1}.

8.4. 1000 Å of silicon dioxide are used in an MOS capacitor. If the silicon is n-type, what is the maximum bias that can be applied in the accumulation mode? The dielectric strength of silicon dioxide is 6×10^8 V m^{-1}.

8.5. The semiconductor doping level in the MOS capacitor of the previous problem was 5×10^{21} m^{-3}. Calculate the capacitance at -2 V.

8.6. Derive an expression for the rate of change of capacitance with gate voltage for an MOS capacitor operating in the depletion mode.

8.7. An MOS capacitor is found to have the following voltage dependence.

Voltage/V	−3	−2	+1.1	+1.2	+1.5	+2.0	+3	+4
Capacitance/pF	200	200	200	180	150	110	100	100

Use the above information to deduce
(a) the conductivity type of the semiconductor
(b) the sign of interface charges.

8.8. Reference to figures 8.4, 8.5 and 8.6 shows that the application of bias bends the bands at the metal-semiconductor interface. What is the magnitude of the bias which must be applied in order to eliminate band bending in a metal-oxide p-type semiconductor structure where the metal work function is 4.8 eV and the semiconductor electron affinity is 4.2 eV?

A metal-oxide n-type semiconductor capacitor has a zero bias value of 1000 pF cm^{-2} and a flat band voltage (V_{FB}) of 0.55 V. If a processing step introduced 15×10^{-9} coulombs cm^{-2} of positive interface charge, estimate the revised value of flat band voltage.

8.9. When an MOS device is operating in the depletion mode, the semiconductor immediately under the gate becomes depleted of carriers. The charge per unit area in this region is $eN_A W$ for a p-type semiconductor. W is the depletion layer width.

The surface potential, ϕ_s, is defined as the difference between the intrinsic energy at the surface, $E_i(s)$, and the intrinsic energy in the semiconductor bulk, $E_i(b)$. The Fermi potential, ϕ_f, is given by

$$\frac{|E_i(b) - E_{f_s}|}{e}$$

It can be seen from figures 8.4, 8.5 and 8.6 that a condition for strong inversion is when $\phi_s = 2\phi_f$.

Once inversion has been reached any increase in bias has no further effect on the depletion layer width. Show that the maximum depletion layer width, W_{max}, can be given by

$$W_{max} = \frac{2}{e} \sqrt{\frac{\epsilon_r \epsilon_0}{N_A} kT \log_e \left(\frac{N_A}{n_i}\right)} \quad \text{for p-type}$$

8.10. The current-voltage ($I_D - V_D$) characteristics in an inversion mode MOSFET can be derived in much the same way as for a JFET (see problem 7.3 in the previous chapter)

$$I_D = \frac{W_0}{L} \mu C_{ox} \left[\left(V_G - V_{fB} - 2\phi_f - \frac{V_D}{2}\right) V_D - \frac{4}{3} \frac{\epsilon_r x_{ox}}{\epsilon_0 W_{max}} \sqrt{\phi_f} \left[(V_D + 2\phi_f)^{3/2} - (2\phi_f)^{3/2}\right] \right]$$

L is the gate length and W_c is the width. C_{ox} is the gate capacitance per unit area.

Use the above equation to develop an expression for the channel resistance of an inversion mode MOSFET when the gate is connected to the drain.

Reference

M. Pepper (1986). *Phys. Bull.*, **37**(1), 22-25; **37**(3), 16-17.

Solutions to Problems

Chapter 1

1.1. (i) 6.04×10^{28} m^{-3}; (ii) $d_{110} = 2.82$ Å, $d_{111} = 2.309$ Å.

1.2. These are (111) planes.
$$\text{Separation} = \frac{4 \times 10^{-10}}{\sqrt{3}} = 2.309 \text{ Å}.$$

1.3. $n\lambda = 2d \sin\theta$ where $d_{100} = 4$ Å. Therefore
$$1 = 8 \sin\theta$$
$$\theta = 7.18°$$

1.4. (i) $\dfrac{\pi\sqrt{3}}{8}$ (68 per cent); (ii) on (110) 5.45×10^{18} m^{-2}
on (111) 1.57×10^{18} m^{-2}.

1.5. 1 kg molecular weight = 6.02×10^{26} molecules = 58.45 kg.
1 m^{-3} = 2177 kg. Therefore
$$1 \text{ m}^{-3} = 2.24 \times 10^{28} \text{ molecules}$$

One molecule AB occupies 4.46×10^{-29} m^{-3}.
There are four AB molecules per unit cube which therefore occupies 1.78×10^{-28} m^{-3}. Therefore
$$d_{100} = 5.63 \text{ Å}$$

Use $n\lambda = 2d \sin\theta$ with $\lambda = 1.54$ Å

for $n = 1$ $\theta = 7.86°$
for $n = 2$ $\theta = 15.87°$

For a screen which is 10 cm away from the source of diffraction, the first-order spot will be 1.38 cm above the beam axis. The second-order spot will be 2.84 cm above the axis. The separation is therefore 1.46 cm.

Chapter 2

2.1. (i) 1.52×10^{-8} m
 (ii) 2.52×10^{-10} m
 (iii) 4.55×10^{-15} m.

2.2. $\theta = 33°$.

2.3. $\lambda = 5.4 \times 10^{-12}$ m. separation = 0.048 cm.

2.4. Resolution $\approx 5.4 \times 10^{-12}$ m.

2.5. Resolution $\approx 4.55 \times 10^{-15}$ m.

2.6. $m_{Ni} = 9.75 \times 10^{-26}$ kg $E = \tfrac{1}{2}kT = 2.07 \times 10^{-20}$ J

$$\frac{h}{\lambda} = \sqrt{2m_{Ni}E} \quad \therefore \lambda = 1.04 \times 10^{-11} \text{ m}$$

$$\therefore d = 3.57 \times 10^{-8} \text{ m}.$$

2.7. $y = \dfrac{\pi^2}{3} - 4\left[\cos x - \dfrac{\cos 2x}{4} + \dfrac{\cos 3x}{9} - \dfrac{\cos 4x}{16} + \ldots\right]$

2.8. Use $\dfrac{x}{\sqrt{2}} + \dfrac{jk}{\sqrt{2}} = m$

$\therefore dx = \sqrt{2}\,dm$

and $x^2 = 2m^2 - 2\sqrt{2}\,jkm + k^2$

$$\therefore f(k) = \exp\dfrac{\left(-\dfrac{k^2}{2}\right)}{\sqrt{2}}\left[2\int_{-\infty}^{\infty} m^2 \exp(-m^2)\,dm\right.$$

$$\left. - 2\sqrt{2}\,jk\int_{-\infty}^{\infty} m\exp(-m^2)\,dm + k^2\int_{-\infty}^{\infty} \exp(-m^2)\,dm\right]$$

$$= \sqrt{\dfrac{\pi}{2}}\,(1 + k^2)\exp\left(-\dfrac{k^2}{2}\right).$$

2.9. Velocity of $\alpha = 2.19 \times 10^7$ m s^{-1}
 distance in 10^{-12} s $= 2.19 \times 10^{-5}$ m
 $\lambda = 4.55 \times 10^{-15}$ m [see problem 2.1 (iii)]
 number of wavelengths $= 4.8 \times 10^9$.

2.10. Velocity of electromagnetic radiation $= 2.9979 \times 10^8$ m s^{-1}
 distance in 10^{-12} s $= 2.9979 \times 10^{-4}$ m.

Chapter 3

3.1. (1) $A^2 \int_2^5 \cosh^2 x = A^2 \left[\dfrac{\sinh 2x}{4} + \dfrac{x}{2} \right]_2^5 = 1$

$\therefore A = 0.019$

(2) $\int \psi \psi^* = A^2 \int_{-\infty}^{\infty} x^2 \exp(-2k^2 x^2) = 1$

Hint: $\int_{-\infty}^{\infty} x^2 \exp(-ax^2) = \dfrac{2}{2^2 a} \sqrt{\dfrac{\pi}{a}}$

$A^2 \dfrac{1}{4k^2} \sqrt{\dfrac{\pi}{2k^2}} = 1$

$\therefore A = 2k \left(\dfrac{2k}{\pi} \right)^{1/4}$

3.2. $\dfrac{1}{\psi} [P] \psi = P = \dfrac{\hbar k [\exp j(kx - \omega t) - \exp[-j(kx + \omega t)]]}{[\exp j(kx - \omega t) + \exp[-j(kx + \omega t)]]}$

3.3. $\dfrac{d^2 \psi}{dx^2} + \dfrac{2m}{\hbar^2} (E - V) \psi = 0$

$V = 0 \quad \dfrac{2mE}{\hbar^2} = k^2$

$\dfrac{d^2}{dx^2} [A \cos kx] + k^2 A \cos kx = 0$

3.4. $\dfrac{10^2}{10^{12}} = 4 \exp \left[\dfrac{-2a}{\hbar} \sqrt{2mV} \right]$

$a = 0.35$ Å.

3.5. Since $\delta = 0.066$ mm at 1 MHz, $\mu\sigma = 73.1$
$\delta = 6.6$ μm at 100 MHz

Trans. prob. $= \dfrac{\text{(Transmitted amplitude)}^2}{\text{(Incident amplitude)}^2} = 4 \exp \left[-\dfrac{0.2 \times 10^{-3}}{6.6 \times 10^{-6}} \right]$

$= 2.76 \times 10^{-13}$

$\dfrac{\text{Transmitted amplitude}}{\text{Incident amplitude}} = 5.2 \times 10^{-7}$

Attenuation $= 10 \log_{10}(5.2 \times 10^{-7}) = -144.6$ dB

3.6. The barriers are at $x = 0$ to $x = a$ and from $x = b$ to $x = a + b$. The wave functions are

(1) $x < 0$ $\psi = A \exp jk_1 x + B \exp -jk_1 x$
(2) $x = 0$ to $x = a$ $\psi = C \exp -k_2 x + D \exp k_2 x$
(3) $x = a$ to $x = b$ $\psi = E \exp jk_1 x + F \exp -jk_1 x$
(4) $x = b$ to $x = a + b$ $\psi = G \exp -k_2 x + H \exp k_2 x$
(5) $x > a + b$ $\psi = I \exp jk_1 x$

The usual boundary conditions are applied to give

$$A = \frac{C}{2}\left(1 + j\frac{k_2}{k_1}\right) + \frac{D}{2}\left(1 - j\frac{k_2}{k_1}\right)$$

$$B = \frac{C}{2}\left(1 + j\frac{k_2}{k_1}\right) + \frac{D}{2}\left(1 + j\frac{k_2}{k_1}\right)$$

$$C = \frac{E}{2}\left(1 - j\frac{k_1}{k_2}\right) \exp[(jk_1 + k_2)a] + \frac{F}{2}\left(1 + j\frac{k_1}{k_2}\right) \exp[(jk_1 - k_2)a]$$

$$D = \frac{E}{2}\left(1 + j\frac{k_1}{k_2}\right) \exp[(jk_1 - k_2)a] + \frac{F}{2}\left(1 - j\frac{k_1}{k_2}\right) \exp[-(jk_1 + k_2)a]$$

$$E = \frac{G}{2}\left(1 + j\frac{k_2}{k_1}\right) \exp[-(jk_1 + k_2)b] + \frac{H}{2}\left(1 - j\frac{k_2}{k_1}\right) \exp[-(jk_1 - k_2)b]$$

$$F = \frac{G}{2}\left(1 - j\frac{k_2}{k_1}\right) \exp[(jk_1 - k_2)b] + \frac{H}{2}\left(1 + j\frac{k_2}{k_1}\right) \exp[(jk_1 + k_2)b]$$

$$G = \frac{I}{2}\left(1 - j\frac{k_1}{k_2}\right) \exp[(jk_1 + k_2)(a+b)]$$

$$H = \frac{I}{2}\left(1 + j\frac{k_1}{k_2}\right) \exp[(jk_1 - k_2)(a+b)]$$

These may be back-substituted to give $\psi = E \exp jk_1 x + F \exp -jk_1 x$ in terms of I, and $|II^*| / |AA^*|$.

Note: The derivation could also have been undertaken to give the wave function in the interbarrier region in terms of A, the incident amplitude.

3.7. Divide problem into three regions
Region 1 $x > a$ $V = 0$
Region 2 $-a \leqslant x \leqslant a$ $V = -V_0$
Region 3 $x < -a$ $V = 0$
It is assumed that particle approaches from $-\infty$.

Solution for region 1 is $A \exp jk_1 x$

$$k_1 = \frac{h}{2\pi} \sqrt{2mE}$$

Solution for region 2 is $B \exp jk_2 x + C \exp -jk_2 x$

$$k_2 = \frac{h}{2\pi}\sqrt{2m(E-V_0)}$$

Applying the usual boundary conditions for the conservation of matter and momentum at $x = a$ gives

$$B = \frac{A}{2}\left(1 + \frac{k_1}{k_2}\right)\exp j(k_1 - k_2)a$$

$$C = \frac{A}{2}\left(1 - \frac{k_1}{k_2}\right)\exp j(k_1 + k_2)a$$

Solution for region 3 is $D \exp jk_1 x + E \exp(-jk_1 x)$.
Applying similar boundary conditions at $x = -a$ gives

$$D \exp(-jk_1 a) + E \exp jk_1 a = B \exp(-jk_2 a) + C \exp jk_2 a$$

and

$$D \exp(-jk_1 a) - E \exp jk_1 a = \frac{k_2}{k_1}[B \exp(-jk_2 a) - C \exp jk_2 a]$$

D and E can be solved in terms of B and C and hence in terms of A. For this problem we are only interested in D.

$$D\left[\frac{A}{4}\exp jk_1 a\right]^{-1}$$

$$= \left(1 + \frac{k_2}{k_1}\right)\left(1 + \frac{k_1}{k_2}\right)\exp(-2jk_2 a) + \left(1 - \frac{k_2}{k_1}\right)\left(1 - \frac{k_1}{k_2}\right)\exp 2jk_2 a$$

A transmission coefficient $|AA^*|/|DD^*|$ can be given in terms of the sine and cosine definitions of the exponentials

$$T = |AA^*|/|DD^*|$$

$$= \left[\cos^2 2k_2 a + \frac{1}{4}\left(\frac{k_1}{k_2} + \frac{k_2}{k_1}\right)^2 \sin^2 2k_2 a\right]^{-1}$$

since $\cos^2 2k_2 a = 1 - \sin^2 2k_2 a$

and since $\left(\frac{k_1}{k_2} + \frac{k_2}{k_1}\right)^2 - 4 = \left(\frac{k_1}{k_2} - \frac{k_2}{k_1}\right)^2$

$$T = \left[1 + \frac{1}{4}\left(\frac{k_1}{k_2} - \frac{k_2}{k_1}\right)^2 \sin^2 2k_2 a\right]^{-1}$$

If $k_1 = k_2$ $T = 1$.
If $k_1 \neq k_2$ T is always less than unity.

3.8. The approach is exactly as per the text.

It is assumed that $\psi_1 = \sqrt{\dfrac{2}{a}} \sin \dfrac{n\pi x}{a}$

$\psi_2 = \sqrt{\dfrac{2}{2a}} \sin \dfrac{n\pi x}{2a}$

are both normalised wave functions

At minimum energy $(n = 1)$

$(H_{11} - E)(H_{22} - E) = (H_{21} - ES_{21})(H_{12} - ES_{12})$

It can be shown that $S_{12} = \displaystyle\int_0^{3a} \psi_1 \psi_2 \, dx = \int_0^{3a} \psi_2 \psi_1 \, dx = S_{21} = S$

$\therefore (H_{11}H_{22} - H_{21}H_{12}) - E\,[(H_{11} + H_{22}) - 2S] + E^2(1 - S^2) = 0$

This quadratic equation will have two solutions.
$H_{11}, H_{22}, H_{21}, H_{12}$ and S can be evaluated in the range $x = 0$ to $3a$ for ψ_1, ψ_2 above.

3.9. $\psi = C_1\psi_1 + C_2\psi_2 + C_3\psi_3$

A treatment identical to that in the text gives

$C_1(H_{11} - ES_{11}) + C_2(H_{12} - ES_{12}) + C_3(H_{13} - ES_{13}) = 0$
$C_1(H_{21} - ES_{21}) + C_2(H_{22} - ES_{22}) + C_3(H_{23} - ES_{23}) = 0$
$C_1(H_{31} - ES_{31}) + C_2(H_{32} - ES_{32}) + C_3(H_{33} - ES_{33}) = 0$

We can assume identical normalised wave functions

$S_{11} = S_{22} = S_{33} = 1,\ S_{12} = S_{21},\ S_{31} = S_{13},\ S_{23} = S_{32} = S_{12}$
$H_{21} = H_{12},\ H_{23} = H_{32} = H_{12},\ H_{13} = H_{31},\ H_{11} = H_{22} = H_{33}$

$\begin{bmatrix} (H_{11} - E) & (H_{12} - ES_{12}) & (H_{13} - ES_{13}) \\ (H_{12} - ES_{12}) & (H_{11} - E) & (H_{12} - ES_{12}) \\ (H_{13} - ES_{13}) & (H_{12} - ES_{12}) & (H_{11} - E) \end{bmatrix} = 0$

E will therefore have three lowest energy solutions.

3.10. $\Delta E = \dfrac{me^4}{8\epsilon_r^2 \epsilon_0^2 h^2} \left[\dfrac{1}{1^2} - \dfrac{1}{\infty^2}\right] = 13.6\text{ eV}$

$\therefore \dfrac{me^4}{8\epsilon_r^2 \epsilon_0^2 h^2} = 13.6\text{ eV} \qquad (\epsilon_r = 1)$

In silicon $m = 0.9 m_e,\ \epsilon_r = 12$

(1) $\therefore \Delta E = \dfrac{0.9}{144}\left(13.6\left[\dfrac{1}{1^2} - \dfrac{1}{4^2}\right]\right) = \underline{0.079\text{ eV}} \quad$ (Bohr orbit)

Note: 3rd excited state has $n = 4$.

(2) $\Delta E = \dfrac{h^2}{8ma^2}\left[4^2 - 1^2\right] = 0.0626$ eV (particle in box)

Note: $a = 2$ (50 Å).

Chapter 4

4.1. $d_{111} = \dfrac{\sqrt{3}a}{3} = 2.5$ Å

$a = 4.3$ Å $= d_{001} \equiv a_z$ (in box)
$d_{110} = 3.04$ Å $\equiv a_y$ (in box)
Distance $(\bar{1}10) - (1\bar{1}0) = 2d_{110} = 6.08$ Å $\equiv a_x$ (in box)
$E = 7.1$ eV

4.2. From density and atomic weight
No. of atoms m$^{-3} = 6.02 \times 10^{28}$
$\therefore N = 3\,(6.02 \times 10^{28}\text{ m}^{-3})$

$E_f = \left[\dfrac{3N}{8\pi}\right]^{2/3}\dfrac{h^2}{2m} = 8.97 \times 10^{-19}$ J $= 5.6$ eV

4.3. The separation between Fermi level and vacuum level is 2.28 eV.
The Fermi level is therefore $5.38 - 2.28$ eV above the ground state.

$N_T = \dfrac{\pi}{2}\left(\dfrac{8m}{h^2}\right)^{3/2}\dfrac{E_f^{3/2}}{3/2} = 2.45 \times 10^{28}$ m^{-3}

Average energy $= 1.86$ eV.

4.4. $P(E) = [\exp(E - E_f)/kT + 1]^{-1} \approx \exp[-(E - E_f)/kT]$

To escape, electrons must have an energy so that

$\exp\left[-\dfrac{4.55e}{kT}\right] \geqslant 10^{-6}$

$T \approx 3825$ Kelvin which is near the melting point of tungsten.
When thorium is added, the probability of escape at 3500 Kelvin is 1.2×10^{-5}.
An oxide-coated emitter has a 10^{-6} probability at 1259 Kelvin.

4.5. $J = AT^2(1 - \rho)\exp\left(-\dfrac{\phi}{kT}\right)$

$1500°\text{C} \equiv 1773$ Kelvin

$10^{-1} = 120\,(1773)^2\,(1 - \rho)\exp\left[\dfrac{-3.55}{8.65 \times 10^{-5} \times 1773}\right]$

$\therefore \rho = 0.969 \equiv 96.9$ per cent reflected

4.6. $J = J_0 \exp\left[\frac{B\mathscr{E}^{1/2}}{kT}\right] \quad B = \frac{e^{3/2}}{4\pi \epsilon_r \epsilon_0}$

$\mathscr{E} = 36 \text{ V m}^{-1}$

4.7. Very low temperatures, therefore all electrons have $E \approx E_f$ or less. An electron will have energy = 5.15 eV outside the metal when $\mathscr{E}x$ = 5.15 eV. Electrons will have 5.15 eV relative to the vacuum level if $\mathscr{E}$ (8 Å) = 5.15 eV or if $\mathscr{E} = 6.4 \times 10^9 \text{ V m}^{-1}$

4.8. $\psi(x + a) = \exp jk(x + a) U_k(x + a)$
$\quad\quad\quad = \exp jk(x + a) U_k(x)$
Since by definition $U_k(x + a) = U_k(x)$
$\psi(x + a) = \exp(jka) \exp(jkx) U_k(x)$
$\quad\quad\quad = \exp(jka) \psi(x)$

Chapter 5

5.1. (a) 9.2×10^{13} m^{-3}; (b) 7.1×10^{19} m^{-3}.

5.2. Lowered from mid band position by 8.62×10^{-4} eV. After addition of aluminium, E_f = 0.056 eV above valence band.

5.3. Use $E_f = \frac{1}{2}(E_d + E_c) + \frac{kT}{2} \log_e [N_d/N_c]$

Substitute this expression for E_f into $n \approx N_d \exp\left[-\frac{E_f - E_c}{kT}\right]$

$n \approx [N_d/N_c]^{-1/2} N_d \exp\left[-\frac{E_c - E_d}{kT}\right]$

$\approx (N_c N_d)^{1/2} \exp\left[-\frac{E_c - E_d}{kT}\right]$

5.4. Material becomes photo-conducting when carriers are excited across the band gap. Therefore

$E_g = \frac{hc}{\lambda} = 1.37 \text{ eV}$

∴ Level is 1.335 eV above valence band.

5.5. Hole concentration = 5.6×10^{21} m^{-3}, $[\mu = V_a/E = 2 \times 10^{-2}$ m^2 V^{-1} s$^{-1}]$.

Diffusivity $D = \mu \frac{kT}{e} = 5.16 \times 10^{-4}$ m^2 s^{-1}.

5.6. $0.5 = (ne\mu)^{-1}$ ∴ $n = 8.32 \times 10^{19}$ m^{-3}

SOLUTIONS TO PROBLEMS 143

$$p_n = \frac{n_i^2}{n} = 2.35 \times 10^{-12} \text{ m}^{-3}$$

5.7. (a) $J = peV_d$
$\therefore V_d = 1248 \text{ m s}^{-1}$
(b) From figure 5.5: at $V_d = 1248$, $E \approx 6 \times 10^4 \text{ V m}^{-1}$
$\mu \approx 2.08 \times 10^{-2} \text{ m}^2 \text{ V}^{-1} \text{ s}^{-1}$
$\therefore \rho = 6.002 \times 10^{-2} \text{ } \Omega \text{ m}$
$\therefore$ Power $= J^2 \rho = 6 \times 10^{10}$ watts m^{-3}.
(c) From figure 5.5: $V_d \approx 2 \times 10^4 \text{ m s}^{-1}$ is about the maximum velocity for which $V_d = \mu E$ holds.
$\therefore J_{max} = peV_d(max) \approx 1.6 \times 10^7 \text{ A m}^{-2}$

5.8. $\rho_s = \rho/t = 0.01/300 \times 10^{-6} = 33.33 \text{ } \Omega$ per square

$$33.33 = 4.532 \frac{V}{I}$$

$\therefore V = 0.735 \text{ mV}$

5.9. Use $p_n(x) = p_{n_0} + [p_n(0) - p_{n_0}] \exp[-x/L_p]$

$[D_p \tau_p]^{1/2} = L_p$ and using $D_p = \frac{kT}{e} \mu_p$

$x = 1.33 \text{ mm}$
This may seem large, but is commonplace with power semiconductor devices.

5.10. $p_n N_d = n_i^2$ $\therefore p_n = 1.96 \times 10^{10} \text{ m}^{-3}$
$\therefore L_p = 2.89 \text{ } \mu\text{m}$.

Chapter 6

6.1. $\phi_B = \frac{kT}{e} \left[\log_e \frac{N_d}{n_i} + \log_e \frac{N_A}{n_i} \right] = 0.716 \text{ eV}$

6.2. $I = I_s \left[\exp\left(\frac{eV}{kT}\right) - 1 \right]$ $\frac{dI}{dV} = \frac{eI_s}{kT} \exp\left(\frac{eV}{kT}\right)$

= slope conductance
$R = 57.4 \text{ } \Omega$ at $V = +0.3 \text{ V}$
$R = 2.57 \times 10^5 \text{ } \Omega$ at $V = +0.01 \text{ V}$

6.3. $\gamma = \left[1 + \frac{\left(\frac{1.5 \times 10^{20}}{100 \times 10^{-6}}\right)}{\left(\frac{10^{24}}{22 \times 10^{-6}}\right)} \right]^{-1} = 0.999\,967$

6.4. Area of single diode = 2.5×10^{-9} m²
∴ Length = 50 μm
The depletion layer (of width W) will be distributed in the p and n sides in the ratio $10^{19} : 10^{20}$.

W is calculated from $C = \dfrac{A\epsilon_r\epsilon_0}{W}$

∴ $W = 8.85$ μm $= x_n + x_p = x_n \left(1 + \dfrac{N_d}{N_A}\right)$

∴ $x_n = 8.04$ μm
Overall length to be occupied by a diode
$= (50 + 8.04 + 5)$ μm $= 63.04 \times 10^{-6}$ m.
Number of diodes per unit area
$= (63.04 \times 10^{-6})^{-2} = 2.52 \times 10^8$ m^{-2}.

6.5. Punch-through if $\left[\dfrac{2\epsilon_r\epsilon_0 (\phi_B + V)}{eN_d}\right]^{1/2} = 10$ μm

(a) V (punch-through) = 6.94 V.
(b) $x_n N_d = x_p N_A$ and $N_A \gg N_d$.
Assume $x_n \approx 10$ μm
∴ $x_p \approx 0.001$ μm
(c) A punch-through diode represents two junction capacitances in series at zero bias.
∴ Capacitance of one junction at zero bias

$= 200$ pF $= \dfrac{A\epsilon_r\epsilon_0}{W}$

where W is the zero bias depletion width.
∴ $A = 5.3 \times 10^{-6}$ m^{-2}

At punch-through the active region represents one capacitor with plate separation of 10 μm.

∴ $C = \dfrac{5.3 \times 10^{-6} \times 12 \times 8.854 \times 10^{-12}}{10 \times 10^{-6}} = 56$ pF

6.6. $\phi_{p^+n} = 0.8$ eV

∴ $W \approx \sqrt{\dfrac{2\epsilon_r\epsilon_0\, \phi_{p^+n}}{eN_d}} \approx 1$μm $= X_n(V = 0)$

$\phi_{np^-} = 0.51$ eV

$W_{np^-} = \sqrt{\dfrac{2\epsilon_r\epsilon_0 (\phi_{np^-} + V_R)(N_d + N_{A^-})}{eN_d N_{A^-}}} = X_n + X_{p^-}$

$$= X_n \left(1 + \frac{N_d}{N_A-}\right) \text{ since } X_p - N_A - = X_n N_d$$

$X_n = 1$ μm when $V_R = 8.8$ V.

6.7. $E_{ABD} = \dfrac{V_{ABD}}{W(V_{ABD})} \approx \dfrac{V_{ABD}}{\sqrt{\dfrac{2\epsilon_r \epsilon_0 V_{ABD}}{eN_d}}} \qquad \phi_B \ll V_{ABD}$

$\therefore V_{ABD} = 11\,934$ V

$\therefore M = \left[1 - \left(\dfrac{V_{PT}}{V_{ABD}}\right)^3\right]^{-1} \approx 1$ since $V_{PT} = 6.94$ V

6.8. $I = I_s \left[\exp\left(\dfrac{eV_f}{kT}\right) - 1\right] \approx I_s \exp\left(\dfrac{eV_f}{kT}\right)$ if $eV_f >$ few kT

$\therefore 1$ volt $= 2V_f + (1000 \,\Omega) I$

$= 2\dfrac{kT}{e} \log_e \dfrac{I}{I_s} + 1000 I$

Equations of this form require an iterative technique for solution. If the diodes were absent the resistor would carry 1 mA. This can be used as a trial value to calculate a new value for I. The process is repeated until the convergent result is obtained.
Start with $I = 10^{-3}$ A in the log of the expression.

$\therefore I_{new} = 10^{-3} \left[1 - 5.06 \times 10^{-2} \log_e \dfrac{10^{-3}}{10^{-5}}\right] = 0.7669$ mA

This value is inserted into the log term

$I'_{new} = 10^{-3} \left[1 - 5.06 \times 10^{-2} \log_e \dfrac{0.7669 \times 10^{-3}}{10^{-5}}\right] = 0.7804$ A

$I''_{new} = 0.779$ mA
$I'''_{new} = 0.7795$ mA, and so on.

Resistor power = 0.607 mW.
Since total power = 0.779 mW
Diode power = 0.086 mW per diode.

6.9. Area = 0.196 cm² $\therefore I_s = 1.96 \times 10^{-5}$

0.75 V $= \dfrac{kT}{e} \log_e \left[\dfrac{I}{I_0} + 1\right] + IR_{series} \approx \dfrac{kT}{e} \log_e \left[\dfrac{I}{I_0}\right] + IR_{series}$

$\therefore R_{series} = 0.041 \,\Omega$

6.10. $\tau = 10^{-6}$ s $Q = 100$ coulombs m^{-3}
This is equivalent to 6.24×10^{20} m^{-3} excess carrier pairs.
Equilibrium hole concentration $p_{n_0} = 1.96 \times 10^{12}$ m^{-3}.
Using $p_n(t) = p_0 \exp(-t/\tau)$
$t = 196$ μs ≈ 200 μs (allowing a margin)
V_{max} must not occur before 200 μs. The time for the first quarter cycle must not be less than 200 μs.
The time for an entire cycle must exceed 800 μs.
∴ Maximum frequency = 1.25 kHz

Chapter 7

7.1. At $V = 0$, depletion width of p$^+$n = 0.63 μm.
∴ $d = 5 + 2(0.63) = 6.26$ μm

$$10 \, \Omega = \frac{0.02 \times \text{(Channel length)}}{\text{(Channel width)} \times \text{(Channel thickness)}}$$

(Channel thickness) = 0.25 μm = $6.26 - 2 W[p^+n(V_g)]$
∴ $W[p^+n(V_g)] = 3.005$ μm
∴ $V_g = 13.02$ V

7.2. In saturation

$$V_D(\text{sat.}) = \frac{eN_d t^2}{8\epsilon_r \epsilon_0} - \phi_B + V_g$$

$$\therefore 0 = \frac{eN_d t^2}{8\epsilon_r \epsilon_0} - 0.6 - 10$$

$t = 7.5$ μm = 10 μm $- 2x_j$
∴ Junction depth $x_j = 1.25$ μm
The two diffused gate components represent two capacitors in parallel.
The capacitance of one gate is therefore

$$10 \text{ pF} = \frac{A\epsilon_r\epsilon_0}{W} = \frac{A\epsilon_r\epsilon_0}{(t/2)}$$

∴ Area = 3.53×10^{-7} m^2

7.5. $\gamma = 0.97 = \dfrac{I_p}{I_p + I_n}$ (pnp)

$\alpha = 0.95 = \gamma B$

$\therefore I_E = \dfrac{I_c}{B} = I_p + I_n$

$\therefore I_p = 9.9$ mA

SOLUTIONS TO PROBLEMS 147

7.6. $\beta = 1000$ $\therefore \alpha = 0.999$

$$B \approx 1 - \left(\frac{W_b}{2L_n}\right)^2 \ldots \text{(npn)}$$

$$\gamma = \frac{\alpha}{\beta} \approx \left[1 + \frac{\sigma_p}{\sigma_n}\frac{W_b}{L_p}\right]^{-1}$$

$$\therefore \frac{\rho_{\text{base}}}{\rho_{\text{emitter}}} = 25.6$$

7.7. $\gamma \approx \left[1 + \frac{N_A e \mu_p}{N_d e \mu_n}\frac{W_b}{L_p}\right]^{-1} = 0.99975 \ldots \text{(npn)}$

$$B \approx 1 + \left(\frac{W_b}{2L_n}\right)^2 = 0.99996$$

$\therefore \beta = \alpha/(1 - \alpha) = 3.458 = I_c/I_B$

$I_B = 28.91 \ \mu A$

Use Einstein's equation $D_n = \frac{kI}{e}\mu_n = 3.36 \times 10^{-3}$ m² s⁻¹

$$\tau_n = \frac{L_n^2}{D_n} = 1.9 \ \mu s$$

$\therefore Q_B = I_B \tau_n = 55 \ \mu C$.

7.8. Base = $10^{22} = N_A$. n-doping = $10^{24} = N_d$.
$L_p = 80 \ \mu m$. $L_n = \sqrt{D_n \tau_n} = 22 \ \mu m$.

$$\therefore \gamma = \left[1 + \frac{10^{22} \times 2}{10^{24} \times 80}\right]^{-1} = 0.99975$$

$$B = 1 - \left[\frac{2}{2 \times 22}\right]^2 = 0.9979$$

A component of the collector depletion layer (x_p) is located within the p-region of the base.

$$W_{\text{collector}} = \left[\frac{2\epsilon_r \epsilon_0 (\phi_B + V)(N_d + N_A)}{eN_d N_A}\right]^{1/2} = 1.19 \ \mu m \text{ at } V = 10 \text{ V}$$

$$= x_p + x_n$$

$$= x_p \left(1 + \frac{N_A}{N_d}\right) \quad \text{since } x_n N_d = x_p N_A$$

$\therefore x_p = 1.078 \ \mu m$

$\therefore W_b = 2 - x_p = 0.922 \ \mu m$ at 10 V reverse bias.

Chapter 8

8.1. $25°C \quad R(+0.1 \text{ V}) = 1.74 \times 10^2 \text{ }\Omega\text{ m}^2$
$R(-10 \text{ V}) = 8.42 \times 10^3 \text{ }\Omega\text{ m}^2$
$125°C \quad R(+0.1 \text{ V}) = 0.377 \text{ }\Omega\text{ m}^2$
$R(-10 \text{ V}) = 6.58 \text{ }\Omega\text{ m}^2$

8.2. $N_d = 10^{20}\text{ m}^{-3} \quad W(V=0) = 2.57 \text{ }\mu\text{m}$
$N_d = 10^{26}\text{ m}^{-3} \quad W(V=0) = 25.7 \text{ Å}$

8.3. $\phi_B = 0.6 \text{ eV}$
$0.01 = (N_A e \mu_p)^{-1} \quad \therefore N_A = 1.64 \times 10^{22}\text{ m}^{-3}$
$W(2\text{ V}) = 0.458 \text{ }\mu\text{m}.$

8.4. n-type $\therefore$ positive bias for accumulation.

$$\frac{V_{max}}{1000 \times 10^{-10}} = 6 \times 10^8 \text{ V m}^{-1}$$

$\therefore V_{max} = +60 \text{ V}$

8.5. $C_{ox} = \dfrac{\epsilon_r \epsilon_0}{x_{ox}} = 3.01 \times 10^{-4} \text{ F m}^{-2}$

$C = \dfrac{C_{ox}}{1 + \sqrt{\dfrac{2\epsilon_0 \epsilon_{ox}^2 V_G}{eN_d \epsilon_s x_{ox}^2}}} = 9.82 \times 10^{-5} \text{ F m}^{-2}$

8.6. $\dfrac{dC}{dV_G} = -\dfrac{C_{ox}}{2\sqrt{V_G}} \left[\dfrac{\sqrt{\dfrac{2\epsilon_0 \epsilon_{ox}^2}{eN_d \epsilon_s x_{ox}^2}}}{\left[1 + \sqrt{\dfrac{2\epsilon_0 \epsilon_{ox}^2 V_G}{eN_d \epsilon_s x_{ox}^2}}\right]^2}\right]$

8.7. (a) p-type.
(b) Negative.

8.8. $\phi_B = 0.6 \quad V_{fB} = -0.6 \text{ V}$
Q surface state/area $= C_{ox} V$
$V = 0.15$
$\therefore V'_{FB} = 0.55 - 0.15$
$= 0.4 \text{ V}$

8.9. Use $\sqrt{\dfrac{2\epsilon_r \epsilon_0 (2\phi_f)}{eN_A}}$ and $\phi_f = \dfrac{|E_i(B) - E_{fs}|}{e} = \dfrac{kT}{e} \log_e \dfrac{N_A}{n_i}$

8.10. Put $V_G = V_D$

$$\frac{dI_D}{dV_D} = \frac{W_c}{L} \mu C_{ox} \left[(V_D - V_{FB} - 2\phi_f) - 2 \frac{\epsilon_r}{\epsilon_0} \frac{x_{ox}}{W_{max}} \sqrt{\phi_f(V_D + 2\phi_f)} \right]$$

$$R = \left[\frac{dI_D}{dV_D} \right]^{-1}$$

Physical Constants and Conversions from Non-SI Units

1 Ångstrom unit	Å	1×10^{-10} m
1 micron (= 1 micrometre)	μm	1×10^{-6} m
Velocity of light	c	2.9979×10^{8} m s^{-1}
Electron rest mass	m_e	9.109×10^{-31} kg
Electron charge	e	1.602×10^{-19} C
1 electron volt	eV	1.602×10^{-19} J
Permittivity of free space	ϵ_0	8.854×10^{-12} F m^{-1}
Permeability of free space	μ_0	$4\pi \times 10^{-7}$ H m^{-1}
Planck's constant	h	6.625×10^{-34} J s
Rydberg constant	R	8.314 kJ (kmol)$^{-1}$ K^{-1}
Avogadro's constant	N_A	6.02×10^{26} (kmol)$^{-1}$
Boltzmann's constant	k	1.38×10^{-23} J K^{-1}
		8.62×10^{-5} eV K^{-1}
Relative permittivity for silicon	ϵ_{Si}	12
Relative permittivity of silicon dioxide	ϵ_{SiO_2}	3.4
Intrinsic carrier concentration (Si)	n_i (room temperature)	1.4×10^{16} m^{-3}

Index

Accumulation mode 123, 128
Activation energy 53
Aluminium 14, 26, 74, 91, 132
Amplitude-frequency 21, 22
Anthracene 2
Attenuation 49
Avalanche breakdown 104, 105, 107
Avalanche multiplication 103

Barium 58
Barium oxide 75
Barium titanate 74
Base access resistance 118
Base transport factor 115, 116, 117, 119
Beer-Lambert law 35, 39
Bergstresser, T. K. 76
Bipolar transistor 111
Blicher, A. 108
Bloch, F. 65, 76
Bloch function 65, 75
Bloch's theorem 66
Body centred cube 4
Bohr, N. 20
Bohr theory 17, 50
Born 30
Born-Haber 2
Born-Madelung 2
Bragg, William and Lawrence 7
Bragg diffraction 27
Brillouin, L. 76
Brillouin zones 70, 71, 72, 73, 85

Built-in potential 95
Bulk barrier diode 106, 108

Cadmium sulphide 2
Caesium chloride 4
Catalysts 63
Cathode ray tube 58
Close packing 3
Cohen, M. L. 72, 76

Collector access resistance 118
Common base gain 115
Common emitter gain 115
Compensated semiconductor 79, 82
Copper 49
Coulombic attraction 60
Coulomb's law 2
Cumper, C. W. N. 45, 51
Current crowding 118

Davisson, C. J. 27
Davisson-Germer experiment 18, 21, 30
de Broglie 18, 26, 30, 36
Debye-Scherrer 8, 20
Delta functions 67
Density of states 55, 57
Depletion layer 93, 95, 105, 106, 117, 123, 133
Depletion mode 124, 129, 133
Depletion mode resistor 130, 131
Devitrification 1

Dielectric strength 133
Diffraction 7
Diffusion 83, 92, 112, 116, 121
Diffusion constant 83, 119
Diffusion current 100, 102
Diffusion length 82, 92, 100, 119
Diffusivity 92
Drain 108, 109, 110
Dulong and Petit law 16, 55

Early effect 117
Eigenfunction 29
Eigenvalue 30
Einstein, A. 76
Einstein equation 64
Einstein relationship 83, 94
Electromagnetic spectrum 7, 8
Electron affinity 120, 132, 133
Error function complement 105
Esaki diode 42
Estermann, L. 20, 27
Exchange integral 47
Extrinsic semiconductor 79

Face centred cube 3, 4
Fermi energy 74
Fermi level 54, 57, 58, 64, 77, 78, 80, 91, 93, 120, 121
Fermi-Dirac distribution 55, 57, 79
Fermi-Dirac statistics 54
Fick's first law of diffusion 83
Field effect transistor 108
Field emission 42, 62
Field emission microscope 63, 64
Field ion microscope 63
Flat band voltage 133
Fourier 27
Fourier analysis 14
Fourier synthesis 28
Fourier's integral theorem 22
Fourier's theorem 21
Fowler, R. H. 76
Fowler-Nordheim tunnelling 62, 75

Gallium arsenide 2, 72, 73, 85, 101
Gate-controlled rectifiers 108
Gaussian distribution 22
Gaussian profile 105
Generation 87
Generation current 100
Germanium 74, 92, 101
Glass state 1

Gold 106, 127, 132
Gunn diode 85

Half-life 88
Hall effect 82, 83, 91
Hall voltage 132
Hamilton, Sir William Rowan 34
Hamiltonian operator 34
Haynes, J. R. 92
Haynes-Schockley 90
Heisenberg 31
Heisenberg Uncertainty Principle 24, 25, 64
Hexagonal close packing 4
Huygen's wave theory of light 18
Hydrogen 50, 51
Hydrogen ions 81
Hydroxyl ions 81

Image charge 60
Image potential 60
Injection efficiency 113, 115, 117, 119
Interface charge 127
Intrinsic concentration 80
Intrinsic Fermi energy 93
Intrinsic semiconductor 78
Inversion 124
Inversion mode 125, 128, 129, 132, 134
Inversion mode resistor 130, 131
Inversion mode transistor 128
Inversion threshold 128
Ionisation potential 51

Kronig, R. de L. 76
Kronig-Penney 65, 67
k-space 23, 70-3

Laplace's equation 109
Latent heat of fusion 1
Latent heat of vaporisation 1
Lattice energy 2
Laue 18
Laue method 8
Lead 88
Lithium 17

Majority carrier 82, 86
Maxwell-Boltzmann 55, 58, 64
Maxwell-Boltzmann distribution 54
Metal-semiconductor diode 122

INDEX

Metastable equilibrium 1
Miller indices 4, 12
Minority carrier 82, 86
Minority carrier lifetime 88, 90, 107, 119, 125
Mobility 83, 90
Molecular orbital 52, 53
Momentum operator 33
Moore, Walter, J. 2, 15
MOS capacitor 127, 133
MOS inverter 130
MOS resistors 130
Muller, E. W. 76
Multiplication factor 107

Napoleon 74
Newton's corpuscular theory 18
Nickel 27, 75
Nordheim, L. W. 76
Normal distribution 22
Normalised wave function 32
NOT gate 130

Ohm's law 85, 92
Overlap integral 47

Palladium 10
Pauli exclusion principle 54
Penney, W. G. 76
Pepper, M. 134
Permeability 49
Petroleum 63
Phosphorus 50, 51
Photoconductivity 58
Photo-electric effect 18, 64
Photolithography 26
Photons 18
Photoresist 26
Pinch-off 128, 129
Planar process 118
Planck 17
Planck's hypothesis 18
Platinum 63, 106
Platinum phthalocyanine 14
Poisson's equation 95
Polyethylene 2
Polymer 26
Position operator 33
Potassium 17, 58
Potassium chloride 11
Punch-through 117
Punch-through diode 106, 108

Quantum Hall effect 132
Quantum mechanical tunnelling 39, 42, 62, 104, 122
Quantum mechanics 132
Quantum numbers 45

Reciprocal lattice 12, 22
Recombination 87, 116, 122, 124
Recombination current 102
Reflection coefficient 38, 59
Relative permittivity 51
Richardson constant 59, 132
Richardson–Dushman constant 59
Ritz–Rydberg formula 17
Rubidium 17
Rutherford 16
Rydberg constant 17

Scharff, Morten 51
Schockley, W. 92
Schottky diode 122, 132
Schottky effect 61
Schrödinger's wave equation 28, 34, 35, 36, 43, 49, 50, 53, 65, 66
Secular equations 47
Sheet resistance 85, 86, 92
Silicon 26, 50, 51, 72, 74, 83, 91, 92, 101, 102, 106, 107, 117, 127, 132, 133
Silicon dioxide 133
Simple cubic structure 4
Single-sided junction 96
Skin depth 49
Skin effect 39
Sodium 15, 17, 58, 74, 127
Sodium chloride 2, 11, 74
Source 108, 109
Space charge neutrality 82
Space charge region 93, 95, 103, 109
Standard deviation 22, 23
Steady-state condition 88
Stern, O. 20, 27
Stern experiment 21
Sulphur 63
Surface breakdown 105
Switched mode power supply unit 122
Switching diode 106
Sze 106

Temperature-compensated voltage reference source 105

Thermionic emission 58
Thompson, G. P. 27
Thompson experiment 20, 30
Thompson's e/m measurement 21
Thorium 59, 74
Thyristor 108
Tin 74
Tin-pest 74
Titanium 132
Transistors 108
Transmission coefficient 42, 49, 50
Triac 108
Triode valve 59
Tsong, T. T. 76
Tungsten 59, 62, 63, 74, 75

Unnormalised wave function 31
Uranium 88

Vacuum level 58, 120
Vanadium pentoxide 74
Varactor diode 98, 127
Variational method 48
von Klitzing 132

Wave function 30
Wave number 29
Work function 58, 64, 121, 132, 133

Zener breakdown 104, 106
Zener diode 42